Calif  U)

D0753059

WITHDRAWN

INFORMATION SYSTEMS IN
LOGISTICS AND TRANSPORTATION

Related Pergamon books

DAGANZO	Fundamentals of Transportation and Traffic Operations
ETTEMA & TIMMERMANS	Activity-Based Approaches to Travel Analysis
HENSHER, KING & OUM	World Transport Research: Proceedings of the 7th World Conference on Transport Research (4 volumes)
LESORT	Transportation and Traffic Theory
ROTHENGATTER & CARBONELL	Traffic and Transport Psychology: Theory and Application
STOPHER & LEE-GOSSELIN	Understanding Travel Behaviour in an Era of Change

Related Pergamon journals

Journal of Air Transport Management
Editor: Rigas Doganis

Transportation Research Part A: Policy and Practice
Editor: Frank Haight

Transportation Research Part B: Methodological
Editor: Frank Haight

Transportation Research Part C: Emerging Technologies
Editor: Stephen Ritchie

Transportation Research Part D: Transport and Environment
Editor: Kenneth Button

Transportation Research Part E: Logistics and Transportation Review
Editor: W. G. Waters II

Transport Policy
Editor: P. B. Goodwin

Free specimen copies available on request.

INFORMATION SYSTEMS IN LOGISTICS AND TRANSPORTATION

Edited by

BERNHARD TILANUS

Eindhoven University of Technology,
The Netherlands

PERGAMON

U.K.   Elsevier Science Ltd, The Boulevard, Langford Lane, Kidlington, Oxford OX5 1GB, U.K.

U.S.A.   Elsevier Science Inc., 660 White Plains Road, Tarrytown, New York 10591-5153, U.S.A.

JAPAN   Elsevier Science Japan, Higashi Azabu 1-chome Building 4F, 1-9-15, Higashi Azabu, Minato-ku, Tokyo 106, Japan

First edition 1997

Library of Congress Cataloging in Publication Data
A catalog record for this book is available from the Library of Congress

British Library Cataloguing in Publication Data
A catalogue record for this book is available from the British Library

ISBN 0-08-043 0546

Printed and bound in Great Britain by Galliard (Printers) Ltd

TABLE OF CONTENTS

FOREWORD

The Center for Transport and Traffic (CTT) is a coordination body at Chalmers University of Technology and Gothenburg University which was formed in 1985 to encourage the acquisition and spread of knowledge required to create and develop advanced systems within the transport sector, such as those concerning the conveyance of goods and passengers, terminal operations, legislation and safety. CTT unites researchers in more than a dozen scientific disciplines in a network for co-operation and exchange of ideas.

Since 1994 CTT has employed professor Bernhard Tilanus, holder of the chair in Quantitative Economic Methods at Eindhoven University of Technology as part-time guest professor, with the task to take part in internationalisation of CTT activities.

Information Technology (IT) has now penetrated society to an extent that would have been difficult to imagine only a decade ago, and has made it possible to apply quantitative methods in sectors of society, which until now mainly have relied on working with pen and paper, voice communication and personal experience.

Gothenburg has since its proud era of East-Indian trade maintained a role as an important hub for export and import of goods to Sweden and other Scandinavian countries. It has therefore been logical for the Board of CTT to take a serious look at the way IT stimulates the use of quantitative methods in distribution systems for goods.

This book is the result of a joint effort of researchers affiliated with CTT and their counterparts in the business environment. It will serve as an important tool for CTT in developing international contacts with universities and transportation industries. It is my hope that in addition it will reach a wide audience of transport professionals, logistics managers, and information specialists as well as researchers and politicians with an interest in the theme of the book.

On behalf of CTT I want to thank Bernhard Tilanus for the energy and dedication with which he has performed his role as an initiator and editor.

Gothenburg, February 1997

Lars Sjöstedt

Chairman, Center for Transport and Traffic
at Chalmers University of Technology and
Gothenburg University

EDITORIAL

This book tries to picture the state of the art in information systems in logistics and transportation, both in research and in practice. It is the product of a joint effort of academicians belonging to the Center for Transport and Traffic (CTT) at Chalmers University of Technology and Gothenburg University, and of practitioners that have many affiliations to the CTT, in the Gothenburg industrial area.

The referees employed by Elsevier Science to advise about the publication of this book, asked: Why Gothenburg? Why not an international authorship to write this book? My answer was threefold.

(1) If the authorship had been international, the book would have been less homogeneous. The authors of this book have a common anchor-ground, the CTT; they have worked together in various combinations, and they have met several times and discussed their contributions to this book.

(2) If this book had been single-authored, say, by a professor sitting in Bordeaux, Bruges or Bologna, the authorship could not possibly have been international, and the cases would most probably have been taken from the author's own country.

(3) Although this book originated from one place on the world map, it is the outlook that counts. It is intended that the issues discussed in this book are relevant for all developed countries at the present time.

The envisaged audience of this book is:

- logistics and information technology managers in large and medium-sized shipping, hauling, and logistics services companies;

- managers of logistics consultancy and software firms;

- transportation economics, logistics and informatics academic staff members and students.

Information systems are interpreted in a wide sense. Information systems include hardware and software, but especially focus on 'orgware': the organizational developments needed to implement newly developed hardware and/or software.

Information systems encompass:

- information technology and telecommunication;

- management information systems;

- decision support systems;

- mathematical models to analyze, simulate or optimize logistics and transportation systems.

Logistics and transportation systems are also interpreted in a wide sense. Logistics and transportation systems cover the entire distribution and supply chain.

From the perspective of the individual shipper in the supply chain, the focus is on external logistics:

- supply, storage and inventory management of input to the firm;

- distribution, storage and inventory management of output from the firm.

From the perspective of the individual haulier in the supply chain, transportation includes all transport modes:

- road;

- rail;

- deepsea, shortsea and inland water;

- air;

- pipeline;

- all combinations of modes, including warehousing and transhipment points within the transportation network;

- packaging, loading and transportation units in the transportation system.

After an introductory chapter discussing the basic concepts that play a part in this book, the main body of the book is partitioned into two parts:

Part I Research directions.

This part describes research actually being done, by young researchers just before or just after their Ph.D. stage, a stage which in Sweden is generally reached after seven years of research, and by some older, managing researchers.

The order of presentation is roughly top-down: the first few chapters deal with broad concepts like sustainability, strategy, marketing, quality, reliability and efficiency; the later chapters deal more with technicalities like data envelopment analysis (DEA), maintenance models and geographic information systems (GIS).

Part II Perspectives from practice.

We are pleased to have found practitioners in the surrounding business world willing to communicate their views on information systems in logistics and transportation. We acknowledge that practitioners have specific barriers to writing papers. Often, interesting material is of a confidential nature. Sometimes, an application is so firm-specific that it is hard to distill a generalizable message. Finally, neither they themselves nor their employers tend to be as keen to publish as academicians are. Nevertheless, information systems in logistics and transportation is an applied topic, and if it is of no use from the perspective of the practitioners, it is of no use at all.

The order of presentation in Part II roughly follows the usual sequence of branches of industry. First come 'real' industries like car manufacturing and bearings manufacturing, then come the service industries, especially the transportation modalities branches, and finally commercial software and research services.

Many thanks are due to Mrs. Jolanda Verkuijlen-Nelissen at Eindhoven University of Technology, and to Mrs. Camilla Hultkrantz and Mr. Ola Hultkrantz at Chalmers University of Technology. Without the activities and the word processing capabilities of these three persons the book would not now be in the reader's hands.

Thanks are also due to the members of the Editorial Board. They stimulated and monitored the production process of this book and refereed the individual contributions. Each contributed paper was refereed by two Board members, and each Board member refereed two or three papers. The final editorial decision was taken by me.

The Editorial Board consists of members of the Board of the Center for Transport and Traffic (CTT), complemented by an international membership. I thank the Editorial Board:

Kenneth Asp, CTT, and VTI, Linköping;

Göran Bergendahl, University of Gothenburg;

Hans Björnsson, CTT, and Chalmers University of Technology;

James Cooper †, Cranfield School of Management, Cranfield, UK;

Nathalie Fabbe-Costes, University of the Mediterranean, Aix-Marseille II, France;

Åke Forsström, CTT, and University of Gothenburg;

Arne Jensen, CTT, and University of Gothenburg;

Kenth Lumsden, CTT, and Chalmers University of Technology;

Lars Nordström, CTT, and University of Gothenburg;

Roland Örtengren, CTT, and Chalmers University of Technology;

Birger Rapp, Linköping University of Technology;

Cees Ruijgrok, Tilburg University, Netherlands;

Lars Sjöstedt, CTT, and Chalmers University of Technology;

Sten Wandel, Linköping University of Technology;

Morgan Williamson, CTT, and University of Gothenburg.

Professor James Cooper participated in an international workshop on information systems in logistics and transportation in Sweden, in January 1996, and contributed stimulating insights.

Soon after, he fell ill and he died in August 1996. His death means a great loss to the international logistics community, from an academic and from an industrial, as well as a personal perspective.

This book is dedicated to the memory of Jim Cooper.

Eindhoven, February 1997

Bernhard Tilanus

1

INTRODUCTION TO INFORMATION SYSTEMS IN LOGISTICS AND TRANSPORTATION

Bernhard Tilanus

ABSTRACT

In this paper, the basic concepts will be discussed which this book is concerned with. In this order, we will discuss logistics (first section), transportation (second section), and information systems (third section).

Keywords: Logistics, transportation, information systems, tracking and tracing

LOGISTICS

Let us start with the entrepreneur. Schumpeter said that the entrepreneur makes new combinations of land, capital and labour, and thereby earns a profit. My grandfather said that entrepreneurship is to turn a dime into eleven cents. The entrepreneur invests, and the return on his investments (ROI) is his profit.

The popular idea is that the entrepreneur invests in plants and machines, i.e. capital goods or fixed assets. Investment analysis is concerned with decision making about this kind of investments. This is indeed the major category of investments in the 'heavy' industry, like the metallurgic or bulk chemicals industries.

In 'light' and highly technical industries, like Philips electrotechnical industries, headquartered in Eindhoven, Netherlands, at least five categories of investments need to be distinguished: (1) capital goods, (2) goodsflows (work-in-progress and inventories of raw materials, component parts and finished products), (3) debts from customers (accounts receivable), (4) research and development (R&D) or innovation, (5) marketing and goodwill. It is increasingly difficult to see each of these categories of expenses as investments; indeed, the last two of them are seldom seen on a balance sheet. But it is good to realise that each of these categories may be equally important, or any of these categories may dominate the others. A firm's ROI may equally be influenced by (1) maintenance and replacement management, (2) goodsflows management, (3) credit management, (4) R&D management, (5) marketing management.

The story goes that just after Worldwar II an American came to Philips in Eindhoven to talk about inventory management. The room was full of researchers from Philips physics laboratories, expecting to be told how to invent things, and there was some disappointment when the talk was only about goodsflows.

This book is concerned with goodsflows management or inventory management, or logistics. So what is logistics?

Webster's dictionary mentions that logistics derives from French logistique, art of calculating, but originally from Greek logos, reason. If the etymological meaning of logistics is reasoning, I am in favour of adopting the wide definition of the Council of Logistics Management (1991):

'Logistics is the process of anticipating customer needs and wants; acquiring the capital, materials, people, technologies, and information necessary to meet those needs and wants; optimizing the goods- or service-producing network to fulfill customer requests; and utilizing the network to fulfill customer requests in a timely way.'

It seems plausible that this wordy definition may be condensed to:

'Logistics is customer-oriented operations management'.

This definition at least enables logisticians or operations managers to engage themselves with the full national economy, not only with the manufacturing industries that account for a mere quarter of a modern Western economy, but also with the service and non-profit sectors that account for the other three quarters.

But for our purposes this definition is too wide. This book is concerned with logistics in the sense of management of goodsflows.

To this end, an older, and more limited, definition of logistics from the Council of Logistics Management will serve, which is also the starting point of the textbook by Lambert and Stock (1993, p. 4):

'Logistics management is the process of planning, implementing and controlling the efficient, cost-effective flow and storage of raw materials, in-process inventory, finished goods, and related information from point-of-origin to point-of-consumption for the purpose of conforming to customer requirements.'

In the sixth edition of their textbook, Johnson and Wood (1996, p. 4) use 'five important key terms':

'Logistics describes the entire process of materials and products moving into, through, and out of a firm. Inbound logistics covers the movement of materials received from suppliers. Materials management describes the movements of materials and components within a firm. Physical distribution refers to the movement of goods outward from the end of the assembly line to the customer. Finally, supply-chain management is somewhat larger than logistics, and it links logistics more directly with the user's total communications network and with the firm's engineering staff.'

Cooper (1994, p. 14) tries to 'encapsulate the widened scope and purpose of logistics management in the 1990s' by the following definition:

'Logistics is the strategic management of movement, storage and information relating to materials, parts and finished goods in supply chains, through the stages of procurement, work-in-progress and final distribution. Its overall goal is to contribute to maximum current and future profitability through the cost effective fulfilment of customer orders.'

This complies with what Cooper et al. (1994, p. 2) say about it:

Logistics 'refers essentially to the management of supply chains in commerce and industry'. They divide logistics management into 'three constituent elements; namely, procurement logistics, production logistics and distribution logistics'.

Two remarks about these definitions:

(1) It is good to realize that there are not only 'customers' at the end of the supply chain, but there are commercial customers all along the way (at each stage where the goods change ownership) whose 'requirements' must be satisfied.

(2) One might say that the 'procurement logistics' of one link in the supply chain is the 'distribution logistics' of the preceding link. Therefore, in this text, we combine the two and only distinguish between external logistics and internal logistics (i.e., production logistics).

Hence, two kinds of goodsflows may be distinguished:

(1) internal goodsflows within a plant, i.e. inventories of raw materials and parts, work in progress with inventories on the way, and inventories of finished products within the plant; likewise, internal transportation is transportation within the plant, typically by lines, chains and other fixed installations; internal logistics is management of these goodsflows; this is the prime meaning given to logistics in assembly firms like Philips industries and in part of the European continent;

(2) external logistics between plants and to the consumers, i.e. collection, transportation and distribution of goods through public space: by road, rail, inland water, shortsea, deepsea, pipeline or air; this is the prime meaning given to logistics in the Anglo-Saxon world.

This book is concerned with the latter kind of logistics.

TRANSPORTATION

Three things may happen to goods in their flow, i.e., we distinguish three kinds of transformations in goodsflows:

(1) They may be changed in form, processed on their way to end products, transformed in form, or transformed in a strict sense;

(2) They may be stored, withheld to be put back into movement at a later moment, transformed in time, or translatated in a specific sense;

(3) They may be physically moved, brought from one place to another, transformed in space, or transported in the normal sense.

We encounter all three kinds of transformation both in internal logistics and in external logistics. However, their order of importance is reversed. In internal logistics the order is:

(1) transformation;

(2) translatation (inventories of raw materials, parts, and finished products);

(3) transportation (physical movement by fixed installations like lines or chains, or by mobile devices like hand trucks or fork trucks).

Whereas in external logistics the order is:

(1) transportation (through public space, mostly by mobile devices like trucks, trains or ships, but also by fixed installations like pipelines);

(2) translatation (inventories on the way, sometimes viewed in the perspective of the whole distribution channel, so including inventories at wholesalers and retailers);

(3) transformation (postponed production in the distribution channel, value added logistics, customizing end products, packing and unpacking, sticking price labels on, etc.).

The three kinds of transformations are usually represented in schemes as follows:

(1) a transformation in form, by a rectangular box;

(2) a transformation in time, by a triangle;

(3) a transformation in space, by a line, where an arrow indicates the direction of the movement if it is other than top-down or from left to right.

Thus, we may depict the supply chain as in Figure 1.

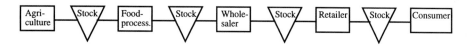

Figure 1. Supply chain from farmer to consumer, from sand to hand, from manure to mouth, etc.

The supply chain refers to the physical flow of goods. Quite rightly, some economists remarked that the supply chain is not supply-pushed, but demand-pulled, hence demand chain is a better name. Economists often use the scheme of the branch column, as in Figure 2.

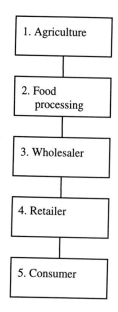

Figure 2. Branch column from farmer to consumer

The branch column refers to the commercial flow of goods. Between the processing firms of different branches are markets. The vertical lines represent trade transactions.

Naturally, the supply chain and the branch column are simplifications. Goodsflows may be split and proceed to different industries (diverge), or they may be combined to form a joint product (converge). Thus we may get complicated pictures of production and distribution networks, which may even have cycles in them (input-output analysis is based on the idea of mutual deliveries between branches, including both 'forward' deliveries in their 'natural' order, and 'backward' deliveries).

Let us take just one link from such a production and distribution network, where the nodes are consecutive processing firms, and zoom in to the relation between the two nodes, which we will call supplier and consignee, or demander. We may distinguish at least four flows in the relation between supplier and consignee, see Figure 3.

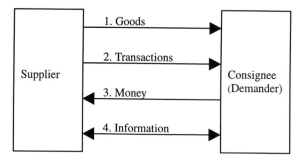

Figure 3. One link in a production and distribution network dissolved into four kinds of flows

(1) Goods flows; are the external logistics goods flows, where transportation is predominant, but where translatation (storage) or transformation (value added logistics) may occur on the way. Depending on the question of who manages which part of the supply chain, the goods flows may be part of physical distribution by the supplier, or part of physical collection by the demander.

(2) Transaction flows; this is a generalisation of the markets in the branch column. From the perspective of the supplier, this is commercial distribution, as opposed to physical distribution; from the perspective of the demander, this is commercial supply, as opposed to physical supply. Often a transaction flow can be broken down into a complicated chain of outsourcings and responsibilities, e.g. see Figure 4; this leads into the field of transportation law and outside the scope of this book.

(3) Money flows (please refer to Figure 3 again); this is mainly payment for the goods received. Between all the flows distinguished between supplier and demander, there are time relationships. If the goods delivery precedes their payment, the supplier invests in the debt of the consignee. However, goods may also be paid in advance. And money flows may be reversed, e.g. if the supplier pays for any damages claimed by the consignee.

(4) Information flows; these go both ways and usually there are a number of them related to each goods flow. Information flows are also closely related to money flows - in a sense one may say that money flows are often transformed into information flows.

This book is concerned with information systems in logistics and transportation. The information flows as depicted in Figure 3 will be our starting point for a discussion of information systems in the next section.

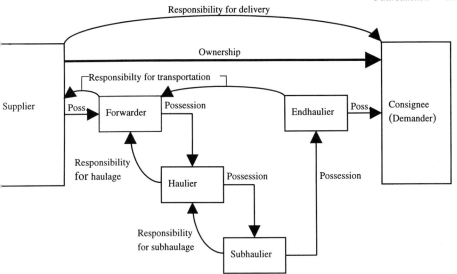

Figure 4. Breakdown of transaction flow; thick straight line: ownership; thin straight lines: possession; curved thin lines: responsibility and liability; insurance companies may enter the picture at each stage

INFORMATION SYSTEMS

We will not here go into the distinction between bits and bytes, data (meaningful bits and bytes), and information (semantic or pragmatic meaning derived from data by a human being). We will use the terms data and information interchangeably.

The same three kinds of transformation that may happen to goods flows, may also happen to information flows. Information may be transformed in form, in time, or in space. In information flows in external logistics and transportation, the order of importance of the various kinds of transformation is from last to first, so we have:

(1) Transformation in space; transportation of information is often called telecommunication. If the telecommunication is two-way between computers belonging to different information systems, we call it electronic data interchange (EDI). EDI plays a major part in this book.

(2) Transformation in time; translatation of information is often called information storage and retrieval. Retrieving information from a computer is far more difficult than storing it. Note that storage of goods implies three things: (a) actually storing the goods or putting them into storage or handling them in, (b) actually keeping the goods in storage, where the three R's are incurred: Room costs for maintaining the storage space, Rent costs for the money invested in the goods, and Risk costs for the goods to go out of fashion or otherwise lose their economic value, (c) retrieving the goods or picking orders or handling them out. Very similar costs may be distinguished for information storage and retrieval.

(3) Transformation in form; transformation of information in this strict sense is what is usually understood by data processing, computerized decision making, simulation or optimization. The topic of this book, information systems in logistics and transportation, offers the scope, therefore, to also discuss decision modeling in external logistics.

Let us go back to our ingredients for depicting production and distribution networks or, as the case may be, information networks:

(1) boxes, to depict transformation processes, executed by successive suppliers and consignees in the network; these are nodes in the network;

(2) triangles, to depict translatation processes, storage and retrieval of goods or information; these are also nodes in the network;

(3) lines, to depict transportation processes of goods or information, goods or information flows; these are the arcs in the network.

If we use these symbols to make pictures, and indeed use the very words like 'network' and 'flows', we commit a dangerously misleading graphical and verbal modeling error. In external logistics, goods and information are not flows. In external logistics, goods are transported in shipments, information is transported in messages. Shipments and messages are discrete entities. As a rule, there are no actual, but at most virtual, links between the nodes in the network. (This is in contrast to internal logistics, where fixed installation or permanent information connections more often provide physical links between the sender and the receiver.)

A more realistic picture of goods and information 'flows' between a supplier and a consignee may be seen in Figure 5.

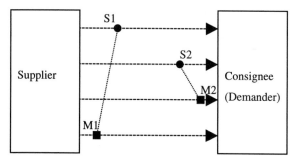

Figure 5. Shipments (S1 and S2), and messages (M1 and M2) in the space between a supplier and a consignee. Broken lines: ideal, virtual routes of shipment or message from sender to receiver; dotted lines: semantic, not physical, relation between a shipment and a corresponding information message. Note: time precedence relations are not represented in the picture

Perhaps we should eradicate the notion of goods 'flows' and information 'flows' from our mind and think of the following definitions:

A shipment is a transmission of a set of one or more discrete handling units, from a shipper to a receiver, that may be handled by human beings or machines; the shipment is forwarded through public space, by various modalities, which may be surface (road, rail or inland water), sea, air or underground, by the own transport means of the shipper or the receiver and/or by professional hauliers in their various modality branches. On their way, shipments may change. They may be combined (consolidated) to larger shipments, or they may be broken down (deconsolidated) into smaller shipments. For that matter, even handling units may change on their way (repackaged into larger or smaller handling units).

A message is a transmission of a set of discrete data, from a sender to a receiver, that may be handled by human beings or computers at either side; the message is forwarded through public space by various modalities, which may be a shipment (the message may be physically connected to a shipment), mail, telephone, telefax, telecommunication (e.g., Internet) or EDI. In external logistics, a message may be semantically or pragmatically related to a shipment.

In external logistics and transportation, a great amount of effort must be spent on maintaining the relation between the information system and the physical goods. Goods identification, tracking and tracing are the key issues. Although there is a lot of overlap between the terms 'tracking' and 'tracing' according to the dictionary, we want to distinguish these terms. Thus, tracking is to find back a shipment when the connection to the information system was broken, to inform any party concerned about the whereabouts of a shipment at his request. Tracing is to follow a shipment as it proceeds through space and to have available information about its whereabouts continuously; this does not necessarily mean that the shipment is 'observed' continuously (for instance, by satellite localisation). At the present time, tracking usually boils down to having a few checkpoints where the shipment should pass, 'observing' the shipment as it passes, and giving a warning message if the shipment does not pass the checkpoint at a predetermined time.

Thus a tracking and tracing system links up an information system with a physical distribution system. The identification is at the level of shipments, or at the level of individual handling units. At one or more (or infinitely many) checkpoints, the tracking and tracing system will:

(1) register the presence, at which time, and possibly the quality status, of the shipment or individual handling units (the Ist-situation);

(2) plan and/or forecast the arrival and departure of the shipment or individual handling unit (the Soll-situation);

(3) compare the Ist- and Soll-situation and take informative action if the two do not match.

Note that (2) and (3) imply that the whole gamut of planning and forecasting transportation operations is included in the tracking and tracing system, but in doing so, its functionality is greatly enhanced.

IN CONCLUSION

This book is concerned with information systems in logistics and transportation. Logistics is external logistics, the management of shipments between shipper and consignee or, depending on the manager's scope of power, the management of the whole or part of a supply chain or a production and distribution network. Transportation is transformation of place: moving shipments through public space by various modalities, over the earth surface, through the air, or underground. Information systems include modeling and management decision making, but key issues are EDI, tracking and tracing.

Readers may be surprised that some articles in this book seem to be rather far off this topic. That is because sometimes a wide perspective is taken as part of a top-down approach. In analogy with Nijkamp et al. (1994, p. 29, 176ff) we view the topic as a prism with five facets, enumerated top-down in the same sequence as the chapters in Part I of this book:

(1) eco ware (the environmental aspects);

(2) fin ware (the financial/economic aspects);

(3) org ware (the organizational aspects);

(4) soft ware (the electronical aspects);

(5) hard ware (the physical, palpable aspects).

REFERENCES

Council of Logistics Management (1991), *Logistics in Service Industries*.

J. Cooper (Ed.) (1994), *Logistics and distribution planning: Strategies for Management*, Second edition, Kogan Page, London.

J. Cooper, M. Browne and M. Peters (1994), *European Logistics: Markets, Management and Strategy*, Second Edition, Blackwell, Oxford.

J.C. Johnson and D.F. Wood (1996), *Contemporary Logistics*, Sixth edition, Prentice-Hall, London.

D.M. Lambert and J.R. Stock (1993), *Strategic Logistics Management*, Third Edition, Irwin, Homewood.

P. Nijkamp, J.M. Vleugel, R. Maggi and I. Masser (1994), *Missing Transport Networks in Europe*, Avebury, Aldershot.

Part I

RESEARCH DIRECTIONS

2

MANAGING SUSTAINABLE MOBILITY: A CONCEPTUAL FRAMEWORK

Lars Sjöstedt

ABSTRACT

As information technology makes it possible to describe and handle increasingly complex transport systems the need to establish a conceptual framework that can serve as a common background for discussion and analysis becomes evident. The multifaceted interaction of transport and its environment makes a conceptual framework a prerequisite for a thorough analysis of such an inherently vague quality as sustainable mobility. This paper is an attempt to provide such a framework.

Keywords: Sustainability, mobility, accessibility, infrastructure

FROM TRANSPORT TO MOBILITY OR ACCESSIBILITY

Transportation has been well established as a scientific discipline at least since the sixties. Originally, traffic was focussed upon. Traffic engineering grew in importance as a profession as most cities found it necessary to employ one or several traffic engineers.

As part of the strong environmental movement and changes of values that swept through universities in 1968 the first strong reactions against unlimited automobilism in cities appeared. This caused a renewed interest in the role of public transport means and partly explains the shift of academic work from traffic to transport that took place in the seventies. While traffic deals with the motion of vehicles along itineraries, i.e. links in a network, transport deals with the movement of passengers and goods from one address to another, i.e. between nodes in the network. To mark their increased customer focus several research institutes and professional associations especially in the US and Great Britain exchanged traffic for transport in their names.

Following the Brundtland report (World Commission on Environment and Development: Our Common Future, Oxford Press, 1987) and the UN conference in Rio de Janeiro on Sustainable Development a further shift has taken place from transport to mobility. The word mobility is frequently used to indicate the ability of a person to undertake movements. Since there is no simple way to measure mobility, it is often taken as synonymous to the actual amount of travel a person makes measured as km per day or per year. However, mobility and travel are from a qualitative point of view quite different concepts. The following example serves to illustrate this. Person A travels 200 km between home and workplace 200 times a year because he cannot find a different job and because his family does not want to move. Person B lives conveniently close to an airport and goes 10 times a year to his weekend house in a sunny warm country with good airline services. They both travel 40 000 km per year but do they have the same mobility? Definitely not. Assuming a travel time per work day of two hours, A has already spent a total of 400 hours travelling, which is above the 365 hours per year that is the accumulated mean of a "natural" travel time budget of 1 hour ± 15 minutes per person per day. Person B has only spent on the average 15 minutes per day and has ample time for engaging in other activities involving travel, should he wish to do so.

A more complete measure of mobility, which includes quality and is frequently employed, is a three-dimensional vector expressing distance covered by travel in passkm/person, time spent by travel in passhours/person and number of trips/person.

The difficulties of handling the concept of mobility of persons becomes even more acute when looking at goods. From an industrial point of view it would, however, make sense if "mobility" of materials, components and finished products could be operationalised.

It can be argued that mobility is not a good term, because it does not reflect the fact that demand for transport is a derived demand; it is always caused by the desire to achieve a different purpose. Therefore it does not make sense to maximise, restrict or minimise mobility per se.

A better term is *accessibility*. It reflects the bottom line: what counts is the possibility to carry out a transport that is fixed in time and space, whenever the need for such a *specific* transport should occur. Thus it is the *ability* to move, not the degree to which this ability is exercised, that characterises a mobile individual as well as a mobile society.

Another advantage with the term accessibility it that it can easily be used and quantified in the context of goods.

WHAT IS SUSTAINABILITY?

Behind the idea of a sustainable transport system lies the general concept of *sustainable development*. Sustainable is a qualitative term. It means that something is compatible with the ecological principles by which nature is ruled. This "something" can be values, attitudes, practices, procedures, systems or products. Thus sustainability can be applied to everything from very soft to very hard notions. It spans from abstract ideas to the real world. In nature sustainable behaviour is mandatory for every organism. This is built into its genes; otherwise it would not survive as a species. The reason that the concept is now also applied to the human race is simple: Although man has been shaped with gifts that make him superior to all other organisms on this planet, ultimately the same biological laws apply to homo sapiens as to all other species.

Large-scale regional ecological catastrophes caused by man are nothing new in human history. The salting of the plains in Mesopotamia and the deforestation of the Mediterranean basin are just two examples. But the massive uses of materials and energy in combination with an exponentially increasing world populaction causes threats of previously unknown dimensions, and the scale is moving from the regional to the global level.

The Brundtland report defined sustainable development as development that satisfies present needs without compromising the capacity of future generations to satisfy their needs.

Sustainable development can also be seen as the one summary answer or solution to the three major problem areas highlighted at the United Nations' Conference on Environment and Development in Rio: *Curbing* population growth; *stopping* the deterioration of the environment; *developing* the economies of the third world. For technology, as cause and cure for many of the problems, this means putting additional emphasis on the social/societal and environmental compatibility of all technical work.

In this paper no attempt will be made to give an operational definition of a sustainable transport system This will be the subject of further research. It may be noted, however, that at least in Sweden the debate on sustainability and transport is strongly linked to the CO_2 issue.

INFORMATION TECHNOLOGY - A PRIME AGENT OF CHANGE

The most important agent of change today is progress in information and communication technologies. The western industrialised countries are now experiencing the full impact of these new technologies. During the last decade many boards of manufacturing companies - much to their surprise - have experienced that it has been possible to lay off between 30 and 40 % of the work force and dramatically cut capital requirements in terms of floor space and stocks, while simultaneously increasing production capacity and vastly improving service to customers.

It is ironic that while this is happening one can still read articles by journalists who claim that introduction of computers does not produce any rationalisation effects. The reason is of course that both introductions of new technologies and the changes they produce occur in an evolutionary way. Thus what is no less than a revolution, when summarised in a historic perspective, becomes just rapid but imprecise change when you live in the middle of it.

The importance of technological change is often debated. Quoting the US economist Kenneth Small it may be said that "in looking ten to fifteen years to the future, analyses of behaviour and of technology tend to merge". Schumpeter argued that technological innovations are the ultimate sources of economic progress. Later research, e.g. by the 1993 Nobel laureates Robert W. Vogel and Cecil North, has shown that organisational innovations are equally important and that lack of a sufficiently stable institutional framework may cause a slow-down of the economy. Fogel claims that only 3 % of the growth in GNP in the US during the great railway era of last century may be explained by access to railway technology as such.

To a certain extent this is a discussion of the chicken and the egg. It is true that while technical innovation is frequently based on some technical invention, it is harder to trace social innovations back to social inventions while maintaining a clear distinction between the two concepts. On the other hand it is likely true, that if one technical innovation would not have been available as a base for economic development, it would have been substituted for a next best technology which could have been used for an almost equally rapid economic development. It is also true that today the options for choosing among separate technical systems and the complexities of these are so manifold that in reality the social values and human preferences will dictate the realisation of the technical contents of future systems.

Irrespective of the extent to which it produces economic growth, information technology does leave its fingerprints on most of the phenomena we observe in our present society. To better understand what is happening in transportation it may be useful to recapitulate the story of information technology from the perspective of a transportation analyst.

Development of information technology during the post-war period may be roughly divided into four periods:

1.   In the fifties and early sixties much of the fundamental applied semiconductor research was carried out while simultaneously military applications and some early university installations were developed.

2.   In the late sixties and seventies, mainly through IBM, civilian data processing was established in separate facilities. These contained sensitive equipment that had to be placed in carefully air-conditioned rooms and maintained by specialists. Digital communication was yet only to a very limited extent employed in the civilian sector.

3.   In the eighties data processing moved to the desk of everybody by the appearance of the personal computer. Huge investments were made in digital communication systems, and large, geographically dispersed information systems were developed. This gave an advantage to the manufacturing industries in relation to the transport industry. As computers were still stationary and communication protocols not standardised, transport industries could only to a limited degree take advantage of the development. This reinforced the fragmentation of this industry. Established opinions of transport systems as a common resource, equally accessible to anyone, were dispirited. A giant wave of privatisation rolled over the continents, and the opinion gained strength that it is up to the individual and the single company to arrange their transport by acting in a market where transport is increasingly carried out by privately owned companies.

4.   In the nineties semi-conductor technology has become truly mobile. Robust equipment, suitable to function in all types of climate has now become available to everyone at affordable prices. This has led not only to an explosion in the use of mobile telephones; it is evident that a more or less complete elimination of distance as a barrier to human interaction is underway.

The new information technologies no longer handicap the transport sector, since they are almost as easily applied in the field as in the office. Instead the transport sector has been brought to the foreground as one of the few remaining huge markets to be exploited by the information technology industry. It is not surprising that we have witnessed a renewed interest in infrastructure issues. After a long period of neglecting long-term transport development, nations once more assume an increased responsibility for infrastructure development. Plans for vast investments have been made in Europe as part of the European integration efforts and support to the former Eastern block of nations. There is also a growing awareness and worry among citizens for the environmental burden caused by transports and a growing feeling that the transport system as a whole is functioning sub-optimally.

INTELLIGENT NETWORKING AND INFORMATION LOGISTICS

Why is the logistics discipline growing in importance and applicability? Logistics is often defined as the art of bringing the right amount of the right product to the right place at the right time. In the following an attempt is made to go behind such pragmatic definitions and use a more philosophical approach.

For a long time production was synonymous with a series of manufacturing operations transforming materials from their original form through various intermediate steps to a marketable product. The conventional wisdom was that transport and storage did not add any value to the overall industrial process; they were just necessary evils that were causing costs.

Logistics has changed our perception of production processes. Now manufacturing, transport and storage are all processes or operations that each in its own way creates utility that the product brings to customers, and for which they are paying.

What does this change mean? The realisation of production processes creates trajectories in a domain with two dimensions. One is time and the other represents a continuum from purely abstract imagination of a product, which still does not exist, through various intermediate stages of realisation in the design phase, such as sketches, descriptions, drawings, models and prototypes, through further intermediate stages in the manufacturing phase, such as raw materials, components and preassemblies, to the final finished product. Putting on logistic glasses is equivalent to adding another domain of equal importance: the time - space domain.

Logistics in the formal sense as defined above is impossible without support of information technology. As industries of today practise logistics it is a child of an emerging information society. However, until now there has been little serious consideration of the real possibilities for change that result from a close coupling of the two domains defined above.

Almost unperceived an equivalent change has occurred in information technology. For a long time information technology was equivalent to computing. Who does not remember the old abbreviation EDP? Electronic data processing is equivalent to manufacturing. Now data bases and transmission are playing an ever increasing role. A data base is a storage point for information and data transmission is equivalent to transportation of information.

What has happened is that we have engineered systems that fully contain the same two domains as mentioned above. We have not fully acknowledged this. Only very recently have we started to envision glimpses of the enormous changes to society that will result from dedicated application of logistic principles to information technology, i.e. from fully exploiting the possibilities of closely coupling the two domains to each other.

Figure 1 illustrates these conceptual similarities. The oval and the arrows mark the departure point and the relative shift of interest towards a balanced view of equal importance of all processes.

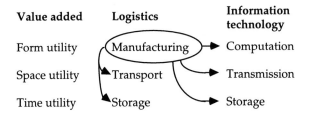

Figure 1. Conceptual similarities between logistics and information technology

One reason that the possibilities are difficult to see is that the language used in information technology is not very clear. A product is the result of a production process. The word information does not make it easy to see the analogies. Information should mean the active process of arranging knowledge in a form that makes it suitable to store or transmit. Instead we are using the word "information" to signify the result of such a process. Thus the word information has become synonymous to the object of itself. This tends to conceal the deeper significance of the word.

One possibility is to coin the word *info*, not as an abbreviation of the word information, but as the object or product of the information process. Info would then be the quantified content of a piece of knowledge, arranged in a form making it suitable for storage in a data base or transmission as a message from one place to the other. Now the equivalence of product and production versus info and information becomes clear and implications easier to realise.

In fact, it is possible to design two models of transportation and communication systems, respectively, in such a way that they completely match each other as shown in Figure 2.

If a consumer needs a product he either buys it off the shelf, or - as is becoming increasingly common in a logistic society - orders it. To be able to place an order the customer first inspects a virtual version of the product, which is available as part of the marketing activity of the producer.

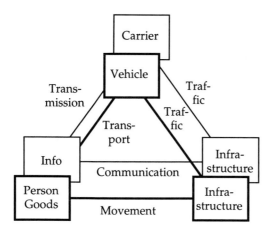

Figure 2. Two models highlighting the similarities between logistics and information technology

This customer inspection frequently requires some form of transmission of an info from a data base. Two observations may be made. The first is that this is already an example of clever coupling of the two domains. The second is that a customer does normally not worry about how the production process works that will deliver the product he has ordered. It is the responsibility of the industry concerned to bring the product to him at the time and the location that has been agreed upon.

It is easy to transform this to a situation of pure information technology. Here is the skeleton of a scenario: A customer has the need of an info. His computer terminal serves both as order entry point and delivery point of the info he is requesting. As result of competitive sales efforts he has free access to virtual versions of all available infos. Once he has placed an order for a specific info, the information process is initiated. This may involve data processing in computers at various locations, temporary storage of semi-finished infos and several intermediate transmissions, before the final info is transmitted to the terminal of the customer and presented to him. It is again typical, that while a customer may need to have some knowledge of how and where to find virtual versions of an info, he has no or little knowledge of how the information process is working.

There is a difference, though: the information system is about a logistic system (cf. Figure 2), whereas the information system about an information system is the same information system.

For a scientist the difference between virtual and real versions of an info may be a complete set of references that ties the info to the body of scientific knowledge.

Information systems that function in this way must possess ability to perform intelligent networking and those who engineer such systems must have thorough understanding of the principles of information logistics.

A GENERAL SYSTEMS MODEL

The models in Figure 2 are not explained in detail here. They have pointed the way to a more recent model, which was developed by Ridley and Sjöstedt (Sjöstedt, 1996). It is shown in its basic form in Figure 3. Instead of movement, the two concepts of accessibility and land use have been introduced, which makes the model structure similar to the structure of classic traffic demand forecasting models.

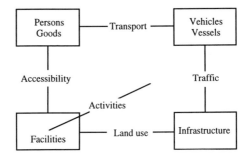

Figure 3.   Systems model of land use, accessibility, transport and traffic

The model consists of four components, each connected to its two adjacent components by four major types of interaction. Some characteristics are briefly described below.

Activities are seen as driving the system but are not strictly part of the model. In Figure 3 this is indicated by showing activities in a dimension perpendicular to the plane of the paper. Although some kind of human participation is normally a key factor of an activity, an activity can in priciple be completely robotised.

Transport infrastructure consists of all manmade permanent facilities, that make movement in the public domain possible. This means that parts of buildings and industrial premises are included, as a minimum the gates or entrances. E.g. an in-door shopping mall is included, even if it is privately built and operated. While most terms lose their sharpness by being given a too broad meaning the opposite is true for the term infrastructure. Transport infrastructure is often given a too limited meaning by referring to facilities for vehicles and vessels, only. The condition that infrastructure is limited to the public domain is customary. However, it is quite possible to apply the conceptual model to an industrial privately-owned premise, and redefine infrastructure accordingly.

Facilities serve as the base for and location of activities. Facilities are normally designed and equipped to allow specific kinds of activities. Although there is much talk about the necessity to provide flexibility in the use of facilities, such as buildings and fixed production machinery, the general trend for facilities is towards increased specialisation. To a large extent this explains the fine web of distinct movements characteristic of modern human life as well as production processes of all kinds. This web is sometimes referred to as an abstract network. The address of a facility belongs to the infrastructure. The entrance to the facility marks the interface between the facility and the infrastructure.

Persons and goods are the participants of the activities. They need to be moved to the location of the facility where their next activity is scheduled to take place. Thus they have a dual role:

in relation to their participation of an activity they are subjects; in relation to their need for movement they are normally objects. Sometimes, however, they assume a subjective role also in the latter case. A pedestrian, a bicyclist and a driver of a private automobile have combined roles as objects and subjects, which gives the explanation why they in a certain sense are perceived as both efficient and effective.

Vehicles and vessels are the means by which the need for movement is satisfied. They are the operating *tools*. Traditionally the choice of transport means limits the scope of the system under study. This is why the whole system is often referred to as the transport system and the academic subject as transportation. Switching to logistics as the academic subject means that focus is shifted from a structural perspective centered around vehicles/vessels and infrastructure to a dynamic perspective focussed around accessibility. Conceptually there is a close relationship between studies of accessibility and supply chain management.

Land use earmarks certain facilities at certain locations in relation to the infrastructure for certain activities. Thus land use is described in terms of locations A, B, C, ... of specific types of facilities and the infrastructure that serves to connect these facilities with each other through networks for transport, communications, energy supply etc.

Accessibility measures the ability of a person positioned at address A to move to another address B. Thus accessibility closely parallels mobility. It is also similar to *movement*. All three concepts are difficult to measure in an unambiguous way. Although this is not common, the concept of accessibility can easily be extended to cover possibilities to move any kind of subject matter from A to B.

Transport performs a change of position of persons or goods. The positions before and after a transport should be identifiable as locations with addresses A and B, respectively. A and B are normally modelled as two nodes in a network. A change of position which does not result in a change of location is called handling. The word transport normally refers to the activity by which the object is transported but it is sometimes used to denote the transported object itself. This latter meaning is synonymous to *load*. In the interest of transport *efficiency*, a load is normally *consolidated*, i.e. consists of several shipments/travellers. The unit of transport is a vehicle or a vessel.

Traffic is the movement of vehicles or vessels along a route connecting two locations A and B. A route is normally modelled as one or several links in a network. The units and route are normally adapted to each other. Together they form a traffic system.

The separate vehicle or vessel component is sometimes skipped as its functions are built into the transport object itself. Thus a pedestrian on a path in nature is a transport/traffic system. Also, goods flows through pipelines or logs transported on rivers in Scandinavia are transport/traffic systems.

THE OPERATIONAL PERSPECTIVE

The basic model of Figure 3 can be seen in several different professional perspectives. One is an *operational* perspective, where the four interactions are seen as *processes*. Here the focus lies on understanding how the processes work and evaluating their performance in terms of efficiency, service provided, cost etc. This is clearly a basic perspective in any analysis aimed at assessing the sustainability of the system. The model is however not symmetric in terms of the four processes. The true dynamics of the system is concentrated in the traffic process, where time constants typically are much shorter than in the other processes, as shown in Table 1.

Table 1. Comparison of typical time constants for action within the four processes of the systems model

Process	Order of time constant
Land use	Decades-years
Accessibility	Years-months
Transport	Weeks-days-hours
Traffic	Hours-minutes-seconds

Significant for the quicker pace of the traffic process is that this is also where the systems metabolism is most intense. Although the other processes also require energy, numerous calculations have shown that the traffic process typically uses more than 90 % of the total energy supplied to the system. Note that although this use of energy is allocated to transport of persons or goods, it belongs to the traffic process. This is also where it may be directly influenced. Thus in any analysis of sustainability of transport systems the traffic element must be analysed in detail. Efficiency of power plants of vehicles in their conversion of energy bound in fuels into kinetic energy will always remain a central element.

But energy supplied to the traffic process and the resulting emissions can be allocated to persons and goods carried by the vehicles. This moves focus into the transport process. Since loading and unloading normally require comparatively small amounts of energy, the most important operational factor is the degree of utilisation of the vehicle. Too often only a fraction of the vehicle capacity is used. Ways of improving the utilisation e.g. through better consolidation of goods and ride sharing in private vehicles give immediate results in terms of sharing the environmental impact from a single vehicle among more objects or users.

At the operational level the accessibility and land use processes have a minor impact on the sustainability of the system as a whole.

THE DESIGN PERSPECTIVE

Another important perspective is design. Technology has often been seen as exogenous to the system. When dealing with sustainability it is imperative to treat technology as an endogenous factor offering powerful tools to change systems in desired directions.

With this perspective focus is shifted from the four processes to the four components. The key role here is played by the vehicles. The two most important factors are the performance of their power plants (as in the operational perspective) and the weight of an empty vehicle relative to its load capacity. A reduction of the vehicle weight is very beneficial since it affects the efficiency of the vehicle both when it is running empty and with a load. Persons and goods cannot be "designed" within the system but the influence of their perceptions of accessibility is important. It includes e.g. requested frequency of services which greatly affects shipment sizes for freight vehicles and passenger demand for public vehicles.

Design of facilities is of course important per se but lies outside the system boundary chosen here which only contains the location of facilities.

THE INTEGRATED PLANNING PERSPECTIVE

The perspective which incorporates the most long-term approach is planning. This is the phase where all aspects of the system are open to modifications. By step-wise trading-off gains and losses between components and processes significant improvements of the total system can be reached. This can only be carried out within the framework of an integrated planning as depicted in Figure 4. Although the operational and design perspectives involve separate planning activities, these should be subject to goals, priorities and guidelines established as a result of a coherent integrated planning effort of the kind envisioned here.

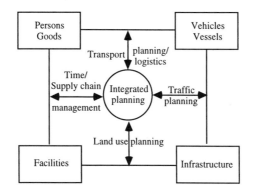

Figure 4. Logistics and planning oriented reference model

Major changes of the system can only be initiated by integrated planning initiatives requiring long times for their full implementation. The investment schemes resulting from land use planning and its associated traffic planning already to a large degree decide the sustainable character of the whole system through their impact on the flow patterns and design characteristics elsewhere.

THE RESEARCH PESPECTIVE

In the academic society freedom of research is highly rated and it is taken for granted that the process of creating new knowledge, which is the task of science, comes first with any potential applications following downstream.

Although basically correct, this view needs to be somewhat modified when dealing with complicated transport systems. The successive realisation of such systems is governed by planning and design efforts that should be led by stable, easily understandable and quantifiable political goals. How to achieve this is an important object of political and social research by itself.

The role of research should be seen as a parallel effort that picks up and in a scientific language problemizes and independently solves the knowledge development needs that arise throughout the policy making and systems development processes. In a way this is creating a pattern where policy making comes first and research last. This should not be confused with lack of freedom of research. On the contrary, it becomes even more important that the scientist is given complete freedom in the process of converting research needs into scientific problems in the context of research methodologies that he or she is able to master. However, considering the grave dangers of research paid for by interested third parties, the scientific community must always exercise the right and the obligation to denounce misuse of this freedom.

CONCLUSIONS

What is new in the application of conceptual models, like those suggested in Figures 3 and 4, and the concurrent interactive activities of policy making, systems development and research, as compared to traditional, disciplinary research and its applications, is that research money will not flow directly into disciplines but rather be tied to the need to create knowledge improving the ability of the funder to solve specific application problems. The work on this application problem must be carried out in parallel to the scientific work of the researcher. This will require an intense lateral communication process, again at the same time stimulated by and requiring the support by advanced information technologies. The overall effort will have to cope with and bridge the differences in time horizons indicated in Table 1 and the hitherto professionally separate planning processes identified in Figure 4.

Transport, especially goods transport, is still seen by many people as a simple task carried out by people with no or low formal education. This may have been true but will not hold long ahead. Transport is quickly being acknowledged as an important human activity, that will require extensive academic training and disciplinary research. Accessibility manifested as mobility is an intrinsic quality of life. It must remain a postulate that by relevant academic training and research the need for mobility can be made compatible with the need for sustainability.

REFERENCES

This paper summarises work reported in the following four references, which themselves contain more extensive lists of references.

Sjöstedt, L. (1992). "IT/EDI in the Swedish transport sector", The workshop on IT - Computer networking in and between economic sectors - distribution and transport, The OECD directorate for science, technology and industry committee for information, computer and communications policy, Paris, France, 1992. Published in "Research Perspectives on Interorganizational Information Systems", Publications of the Turku School of Economics and Business Administration, Series Discussion and Working Papers 15:1992.

Sjöstedt, L. (1994). Sustainable engineering: The challenge of developing transportation for society, Opportunities in the long term, Published in Proceedings of the 10th CAETS Convocation, Zürich 1993, pp. 337-346.

Sjöstedt, L. (1995). Transportation and Logistics: Towards a Unifying Theoretical Base, School of Technology Management & Economics 1994/1995. Chalmers University of Technology, Gothenburg, pp. 35-48, Sweden.

Sjöstedt, L. (1996). Theoretical Framework: An Applied Engineering Perspective in Mobility, Chapter 7 in Mobility, Transport and Traffic in the perspective of Growth, Competitiveness, Employment, European Council of Applied Sciences and Engineering, Paris.

3

INFORMATION TECHNOLOGY AND DISTRIBUTION STRATEGY

Anna Dubois
Lars-Erik Gadde

ABSTRACT

This chapter deals with the impact of information technology on distribution strategies of manufacturing firms. The potential role of information technology is considerable owing to the fact that it can contribute to rationalization and development of the flow of information as well as the flow of materials. These effects, however, will result only when the potential of IT coincides with the strategic distribution objectives of the manufacturing companies. Therefore, an exploration of current issues in distribution is needed to understand the present status of IT in distribution as well as future developments. The conclusion of the chapter is that so far the effects of IT have been less substantial than was expected ten years ago. However, the analysis shows that IT provides major opportunities for restructuring of distribution arrangements, which means that the long-term consequences might be very significant.

Keywords: Information technology, distribution strategy, distribution systems, differentiated distribution

DISTRIBUTION SYSTEMS AND INFORMATION TECHNOLOGY

A common framework when analysing a distribution system is that it is closing a gap between a production system on one hand and a consumption system on the other. The closing of this gap has been conceptualized in various ways. Some writers provide an illustration of a bridge spanning the gap. Another view is that of the distribution system as handling three flows between production and consumption - a flow of goods, a flow of information and a flow of money. It is rather obvious therefore that contributions from the development of information technology (IT) have been regarded a considerable source for improvements of distribution efficiency. IT should enhance speed as well as accuracy of the flow of information. Furthermore, making use of IT should provide opportunities for improved control of the flow of goods.

Research on distribution systems also confirms that companies are changing their distribution strategies and reorganizing distribution systems today. According to one study "a number of companies have out-stripped their competitors with imaginative strategies" (Stern & Sturdivant 1987). Another author reports that "tremendous innovation in distribution has already begun" (McKenna 1988). In a third article it is argued that "by 2000 industrial distribution will be restructured radically" (Morgan 1991). One reason for the potential magnitude of the on-going changes is that distribution systems historically have been considered conservative and slow to adapt to new conditions (see e.g. Nieschlag 1954 and Stern & Sturdivant 1987).

The aim of this chapter is to analyse the impact on distribution strategies and distribution systems owing to developments in information technology. The role of IT in reorganization and restructuring that has already been undertaken will be illustrated and analyzed. Furthermore potential future impacts will be discussed.

The out-line of the chapter will be as follows. We start by going back to the 1980s to identify the expectations on the potential impact of IT on distribution that were prevailing at that time. After that will follow a brief presentation of the major strategic issues characterizing distribution in the 1990s. These two sections will be necessary to understand the analysis of the IT impact on distribution that can be identified until now. The last sections will consist of a discussion of potential impact on future distribution. This discussion will take the conclusions regarding the present development as the point of departure.

INFORMATION EXCHANGE AND THE EXPECTATIONS ON IT

In the 1980s the potential impact of IT on industrial markets was discussed in numerous articles. To structure our analysis we need some kind of conceptualization of the characteristics of information exchange in a customer-supplier relationship. From a purchasing point-of-view three types of information exchange were identified by Gadde & Håkansson (1993):

- technical information

- commercial information

- administrative information.

The technical information content in buyer-seller communication is related to the technical aspects of a product or a service. Exchange of technical information is concerned with the buyer's need of assistance in problem solving in specific procurement situations. Sometimes a customer will need advice for choosing the specification which will give optimal performance from among the standardized product range of a supplier. This may relate to the choice of size, tensile strength or degree of corrosion resistance. The technical information then regards "what" to buy. It is important thus for a producer to make this kind of information available in one way or another.

Commercial information is information to be used by the customer for the evaluation of various suppliers when a purchasing decision is to be taken. Commercial information is thus used to establish who the exchanging partners in a specific business transaction will be - that is to address the question "who" in a buying-decision process. Commercial information includes conditions regarding price levels, delivery times, etc. of the suppliers that are considered. To become a potential supplier producers therefore must in some way be able to transfer this information to the end-user. This can be done in different ways; through personal information exchange, through price-lists and product brochures or through computer links that can be continually up-dated.

For the selling firm another kind of commercial information can be identified which is expected to have an effect on marketing performance. Information about customer characteristics and buying behaviour can be used for so called data-base marketing. By using data-base marketing a selling firm is supposed to improve the classification of prospective and existing customers and also to develop a lasting association with them (Fletcher 1992).

Administrative information deals with handling the formal exchange between the parties in a specific transaction - i.e. to manage issues regarding "how much, where and when?". To complete a business transaction requires a considerable exchange of information. Orders, documents, invoices, payments etc. have to be transferred between buyer and seller. The paper-flow in a buyer-seller relationship can thus be very substantial. Therefore it constitutes a major potential for rationalization. This is especially important for products characterized by a low unit value, where these administrative costs can account for a considerable part of the total costs.

When IT was introduced, therefore, the expectations on efficiency improvements in distribution were enormous. In the literature it is possible to identify two major directions concerning the potential impact. The first one can be characterized by the following headings from three articles:

- Automation to boost sales and marketing (Moriarty & Swartz 1989)

- IT increases the power of buyers (Porter & Millar 1985)

- The logics of electronic markets (Malone et al. 1989)

The last sentence reflects the underlying thinking within the scope of this direction. By using IT it should be possible to remove the negative effects of existing market systems that were related to insufficient capacity to handle information. IT should provide opportunities to identify all the potential actors in the market - and also to evaluate them in a more or less complete way. Suppliers were offered opportunities in direct marketing and selection of specific customer groups to which more or less individual messages it should be possible to transmit. Customers should now be able to undertake purchasing in a completely rational way by having full information about existing supplier alternatives and the conditions that characterized their offers in every specific buying situation.

The proponents of this view advocated that efficiency is created by choosing the most appropriate partner for each separate transaction - which means to make use of the market forces. By using IT, firms would be better off in this process. One important driving force was thus to identify new business partners. In our terms the major potential impact should be in enhancing the exchange of commercial information.

The second direction, on the other hand, took its point-of-departure in the existing business relationships. Research on customer-supplier relationships on industrial markets has shown that they often are long-term in nature (HŒkansson 1982, Gadde & Mattsson 1987). The advocates of the second direction therefore tried to figure out how IT could be used to make these on-going business transactions more efficient. This direction was not heard and seen as much as the first. What is interesting, however, is that it was presented earlier. In the 1960s ideas were discussed about ways of connecting customers and suppliers closer to each other by "boundary-crossing data systems", thereby accomplishing mutual gains in efficiency (Kaufmann 1966). In a more recent article the benefits of doing so were proposed in the following way (Scott-Morton 1991):

> Electronic sales channels eliminate much of the paper handling and clerical work associated with making a purchase; processing the order, billing the customer, tracking the delivery and accounting for the sale require many people and take a long time. Electronic sales channels stream-line much of that. Some systems let customers reduce their materials inventory by arranging just-in-time deliveries of components. All of this translates either directly or indirectly into savings for the customer.

It is obvious that this direction is related to what we have identified as the exchange of administrative information. According to this school, then, the most significant effects were expected through making existing business relationships more efficient rather than creating new ones.

The potential role of IT in the exchange of technical information was not much discussed in the sense it has been defined here - as a kind of problem-solving activity in purchasing. The major effects of IT regarding exchange of technical information are related to activities in design and construction where substantial improvements were expected and also have been achieved. These aspects will not be taken up in this discussion.

Our analysis is mainly undertaken from a producer perspective. From our discussion follows that producers could expect much from the evolving IT. They should be provided with means to reach new customers and expand business in this way. The same tools could be used also to make on-going business transactions more efficient.

However, one important lesson from earlier research is that IT will have only minor impact on business operations unless it coincides with strategic ambitions that it can support. To be able to identify and analyse the potential opportunities we need thus to take the producers' strategic issues regarding distribution into consideration.

STRATEGIC DISTRIBUTION ISSUES FOR PRODUCERS

At a Swedish symposium dealing with on-going tendencies and future trends in distribution two major producer issues were identified. The first one was the requirement on producers to be able to establish future distribution solutions that will offer the customers value for money - i.e. they should be efficient from a cost-revenue perspective. Over time distribution costs have come to account for an increasing portion of the total cost for a product delivered to an end-user. Production costs have been substantially decreased owing to rationalization efforts and automation providing economies of scale. Costs of distribution on the other hand have increased. One reason is that distribution systems generally have improved their service performance - which obviously leads to increasing costs. Another one is that distribution activities have been difficult to render more effective owing to a lack of rationalization tools. During the 1990s, however, a number of very significant improvements have been attained in distribution. IT is one of the tools that have made these improvements possible.

The second issue identified was that manufacturers felt it very important to get as close to the customer as possible. "Customer" then referred to the end-user of the product. The reason why it has become important to come closer can be explained by the concept of the augmented product. Over time core products have been substantially standardized in many industries. The basis for differentiation and competitive power, therefore, is shifted towards various value-adding activities, such as delivery services, training, after-sales service, etc. That kind of added value is created in the direct contact with the end-user of the product. These contacts are often handled by independent distributors. If these activities become more important from a strategic view-point such an arrangement can be perceived as unsatisfactory. Producers therefore might be interested in getting a direct contact with end-users in some way. Another option would be to establish new forms of relationships with distributors so they no longer are independent with respect to the manufacturers. Information technology can be one way to accomplish both these ambitions. Computer links can make it possible to establish direct contacts with end-users. Furthermore investments in joint information systems can be one way to establish closer relationships with distributors.

The two strategic issues are interrelated. Increasing costs of distribution is not a new problem for producers of industrial goods. Over time producer firms have witnessed expanding costs especially for physical distribution and personal selling. This has made it very difficult to serve small and medium-sized customers in a cost-effective way. In many cases the quantities demanded by these customers have been too small to cover the costs for personal selling of single product lines. By using industrial distributors for serving these kinds of customers, producers have been able to establish cost-effective transactions with these end-users. Distributors can serve small customers more efficiently because their costs for personal selling and physical distribution can be distributed over a huge product range. This is what leads to a functional spin-off (Mallen 1973), which means that distribution activities are taken over by distributors. Such changes over time have led to a situation where independent distributors have come to account for an increasing role in industrial distribution (Webster 1975, Michman 1980, Hutt & Speh 1983, Corey et al. 1989, Gadde 1994).

Introducing intermediaries in industrial distribution solved the cost problem for producers, but gave rise to other problems. Specialization of distribution activities as a result of functional spin-off separated producers from consumers. Already in the 1940s some authors pointed to major problems owing to that kind of structural change. The separation caused a lack of understanding between producer and consumer which especially "raised certain fairly serious problems regarding the quality of the goods made and sold". It also meant "the loss to the producer of practically all control over the conditions under which his product was sold and used"(Alexander et al. 1940). The period from 1940 has been characterized by a substantial increase in the degree of specialization of any industry. Products and processes have become more complex and the need for coordination and integration have increased. At the same time independent distributors have been able to improve their positions in distribution, mainly due to the ability to provide cost advantages in handling small and medium-sized customers. It is not surprising therefore that the problems identified in 1940 are even more accentuated today than they were at that time. These are the reasons for producer ambitions to "get closer to the customer".

IT obviously can contribute to rationalization of costs in the information and material flows. In this way it might help in establishing new kinds of bridges to end-users without the corresponding problems that the bridges used so far have caused.

IT AND IMPROVEMENTS OF DISTRIBUTION PERFORMANCE

So far it is very clear that the major IT-effects in distribution have been connected to rationalization of the exchange of administrative information. We agree very much with the conclusion provided already by Johnstone & Vitale (1988):

> Organizations are using electronic links to form closer links with trading partners which indicates the evolution of electronic hierarchies rather than open competition based on electronic markets.

The most significant IT-impact thus has been to make existing relationships more efficient, rather than to establish new ones. According to more recent findings this state-of-the-art has not changed since 1988 - probably it is even more manifested today. This development is not in line with the expectations from the mid 1980s when exchange of commercial information was supposed to be the primary potential for IT. The reason for this change is very much related to new conditions regarding purchasing. Over the last fifteen years an alternative view of purchasing efficiency has become more and more significant (Carlisle & Parker 1989, Gadde & HŒkansson 1993, Lamming 1993). Buying firms have identified new ways of improving performance by establishing closer collaboration with a reduced supplier base. Major benefits then can be obtained through making adaptations and thus creating dependencies to individual suppliers. These adaptations can take various forms, they can be connected to the product design or the production systems. Also the material flow and the exchange of information have been subject for such adaptations. Owing to these strategic changes it is easy to understand that IT has played a more important role in the exchange of administrative information than in commercial or technical. The two latter types of information exchange therefore will only be briefly discussed.

Technical and commercial information

The information used by customers for technical specification of the item needed can be transmitted in different ways. It can be transferred on a personal basis, through product information brochures, or made available through telephone services. Parts of this information can be standardized and IT-based. One example of such an application is a computerized catalogue developed by SKF, the CADalogue, providing the customer with recommendations concerning suitable ball bearing dimensions. The system suggests a specific type of bearing based upon customer specification of some central parameters regarding the application. In other industries producers have made similar services available on CD-rom or disks.

Commercial information has not been computerized in a way that was expected ten years ago. Anyhow, there exist information systems with the basic idea of handling commercial information exchange. Mostly these systems have the form of a product data-base containing the product range of the various suppliers of specific products. The suppliers connected to such systems up-date their product information and business terms continually. Therefore they present the buying firms with an overview of existing suppliers and the conditions for dealing with them. One example is a Swedish system covering suppliers in construction materials, hospital products and food supplies for canteens and refectories. To be effective for the customer such a data-base has to be linked to a system where orders can be submitted electronically. Without such a connection the data-bases will not meet a significant demand as the major rationalization for customers will be in making transactions more efficient.

We mentioned earlier that data-base marketing has been identified as one possible use of IT for improving sales activities of suppliers. From the selling perspective we have already seen that in business-to-business markets the interest in electronic markets, for good reasons, is limited. But also in consumer markets, applications seem to be rather limited. Fletcher (1992) concludes from a study of data-base marketing of passenger cars that the impact has been rather limited. Few companies seem to have sufficient customer information to profile and segment their customer bases and their data collections or campaigns. In other industries however a number of successful applications have been reported (Verity 1994). It is obvious also that Internet will have a considerable future impact on the flow of commercial information. According to Spar & Bussgang (1996) the Internet 'promises to be the site of a commercial revolution'. However, a conclusion from their article is that 'the Internet has yet to deliver on its promises'.

Other forms of IT have also been used for improvements of marketing performance. One example is concerned with the implementation of expert systems. In an article by Mentzer & Gandhi (1993) three illustrations are presented dealing with advertising, business negotiation and physical distribution. Telemarketing is an alternative way of enhancing effectiveness and efficiency in exchange of information (Voorhees & Coppett 1983). As shown by the authors, telemarketing can increase performance in marketing activities when it is undertaken internally by a manufacturing company. But also specialized companies conducting telemarketing for a number of firms have appeared. We want to emphasize that one of the roles that the authors identified for telemarketing was "selling to marginal accounts". That role, so far, has often been fulfilled by independent distributors. This is therefore an example of a new kind of intermediary in distribution. The same holds for the providers of product data-bases earlier discussed. We will come back to new actors of this type when discussing future distribution structures.

IT and administrative information

It has already been mentioned that the major IT impact on distribution has been reported for the kind of information exchange that is related to the "paper handling and clerical work with making a purchase", in the terms of Scott-Morton (1991) and Kaufman (1966) "boundary-crossing data systems". As was suggested in the predictions of the 1980s IT can make these activities more or less automatic which leads to substantial cost savings especially for business relationships characterized by a large number of transactions. In such cases the number of people occupied by these tasks can be reduced both at the buying and the selling company.

The first IT-systems of this type seem to have been introduced by distributors. One of the most impressive examples was reported from McKesson Drugs - a wholesaler of pharmaceutical products. In the middle of the 1970s McKesson Drugs established an IT-system connecting themselves both with their distributors and the most important suppliers. The number of warehouses could be reduced substantially as well as the people dealing with contacts with suppliers and distributors (Corey 1985). In Sweden two wholesaling companies were innovators within this area. One of the reasons for companies of this type to be pioneers is that they were the dominating suppliers of their resellers. Therefore, the benefits provided by the internal IT-systems developed by these companies could be easily transferred to the retailers by placing computer terminals at their stores, thereby making exchange processes more efficient.

The major effects, however, will be attained when such systems for managing the information flow are combined with IT-systems for control of material flows. The best known - and probably the most significant - is the Odette-system developed by European automotive manufacturers (Figure 1). Odette is an example of a distribution rationalization initiated by a powerful group of customers. The automotive manufacturers jointly decided what kind of information should be exchanged as well as in which way exchange should be undertaken. This way of standardizing exchange routines between buyers and suppliers clearly is beneficial for both parties as long as it can be used for communication with all counterparts. For an analysis of the effects of the Odette-system see Dubois et al. (1996).

There are also many examples reported of supplier efforts to improve efficiency by making use of information technology in this way. Bagchi (1992) provides a number of examples regarding improved performance in international logistic operations of globally active enterprises due to investments in IT. In consumer markets the case of Benetton is a well-known example of an IT-controlled global logistic operation linking together 180 suppliers of raw material, more than 400 manufacturing units and 6000 retailers to customers in more than 80 countries (Bruce 1987).

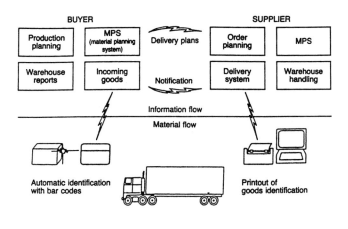

Figure 1. Illustration of the Odette-system (Gadde and Håkansson 1993:162)

A number of multi-national companies have centralized stock-holding to one single inventory location. Atlas Copco Tools is one of the Swedish companies that have withdrawn inventories from the market companies in Europe to one centralized warehouse in Belgium (for a more comprehensive analysis of this case, see Dubois et al. 1996). Earlier the company relied on two central warehouses in Sweden and huge inventories at the European market companies. It is fairly well known that centralizing stocks results in reduced costs. It is more surprising maybe that it also can result in improved service levels. However, for companies with a wide product range it is impossible to keep the whole range in stock at each market company. Therefore, it used to be that most orders could not be delivered in the complete way the customers wanted. At the central warehouse, however, it is possible to store the full product range. By utilizing modern EDI-systems (electronic data interchange) and developing an efficient logistic system it was possible to respond to customers orders in a way that, as an average, was superior to the situation when stocks were closer to the customers. The delivery service was measured as the portion of all orders that could be delivered as requested. When the new system had been established it increased from 70 per cent to 93 per cent. This was accomplished at the same time as the costs were reduced very substantially. The inventories as a proportion of sales decreased from 55 per cent to 18 percent. It should be observed that these effects were obtained also as a result of the fact that the number of production plants successively was reduced from seven (in three different countries) to only one. The experiences of Atlas Copco Tools are a significant illustration of the fact that major effects on total costs can be attained when combined actions to affect production and distribution activities are undertaken.

We have identified examples of IT-systems established by firms on all the three levels of a traditional distribution system. The Odette-system was initiated by a group of customers, the centralized inventory systems (represented by Atlas Copco) have been created by producers and the very first systems of this type were established by existing intermediaries. But also regarding this type of information exchange new forms of intermediaries have appeared. Computer firms like IBM have identified a potential role for a wholesaler in information who connects the information systems of customers and suppliers and works like a switch-board for computer communication. A number of EDI-systems have been applied and others are currently developed that will make similar systems for communication available. It is obvious therefore that new kinds of intermediaries have evolved which specialize in making the information flow more efficient. Sometimes these new firms have been called information brokers. Some of these brokers are using various forms of Internet applications to establish a position in distribution (Harrington & Reed, 1996).

STRATEGIC OPPORTUNITIES FOR PRODUCERS

A number of strategic options for improving distribution efficiency have thus been made available to producers. Development of IT has made it possible to improve speed as well as accuracy in information exchange. Producers have been provided with tools to communicate directly with end-users even when some kind of intermediaries are used. Furthermore these IT-systems increase the potential for control of material flows. New types of intermediaries are established which make entirely new forms of channel arrangements possible.

At the same time producers are challenged by demand from end-users to improve efficiency in distribution operations. Costs for purchased goods and services account nowadays for more than half of total costs in most companies. The buying companies therefore have put an increasing pressure on efficiency in purchasing operations. These requirements in turn will be directly transferred to demands on the distribution system.

The increasing demands on distribution performance will be a challenge for existing distribution systems in most industries. One characteristic of these systems is that they have not been very customer oriented. They have been designed mainly for the sake of getting products efficiently out of the production system rather than efficiently into the consumption system.

Various customers have different perceptions of what should be an efficient distribution solution for them. Some customers might have a rather "simple" supply situation. They will give priority to low cost alternatives and will not be interested in paying for more advanced solutions. To other customers a more sophisticated, tailor-made and more expensive logistical concept can be more effective because it might lead to lower costs and/or improved performance of the manufacturing operations. These firms will consequently be prepared to pay for more advanced solutions.

To successfully serve customers with different demands in an efficient way will make it necessary for manufacturers to develop more differentiated distribution solutions than those existing today. At present customers are handled in a rather general way in most industries and by most firms. The solutions available seem to be based on some kind of average need. Manufacturers continuing to do so will have difficulties in competing with firms that specialize on one of these very different needs. The "average" solution will be problematic to apply when it comes to satisfying both the low-cost demand and the most advanced demands.

Therefore the major conclusion regarding future distribution systems is that there is an urgent need to develop more differentiated distribution solutions. A manufacturer therefore needs to offer standardized distribution solutions as well as more advanced and customer adapted distribution solutions.

Low-cost distribution alternatives

To satisfy the demand for a low-cost channel producers have to look for alternative arrangements. So far independent distributors have been the tool to satisfy this demand. But personnel costs and physical distribution are becoming more expensive also for these firms. Therefore, producers need to consider even more cost effective distribution alternatives. The role fulfilled by distributors has been to handle the flow of material and the flow of information. We have seen that a number of new alternatives have been made available for producers to handle the flow of information. Producers can use EDI to be directly connected to end-users. They can also use some of the new forms of information brokers that have appeared. It is fairly obvious therefore that we can see a tendency towards the emergence of intermediaries specializing on improving performance of different parts of the total flow of information. For customer relationships where the need for exchange of technical information is of minor importance such specialized intermediaries can provide substantial rationalization potential for producers and users.

However, distribution is not only handling a flow of information. Existing intermediaries have taken care also of a flow of goods. The major activities in this material flow are related to transportation and warehousing. Also these activities have to an increasing extent been taken over by specialized firms. Outsourcing of transportation activities to independent service providers is reported by a number of authors (see for example Daugherty & Dršge 1990). Bardi & Tracey (1991) found that more than 75 per cent of the companies in a survey allocated one or more transportation activities to other firms. The most important reasons for the out-sourcing were cost savings in labour and transportation, due to asset reductions and reduced inventories, when transportation was made more efficient. Also logistical planning as a whole had been taken over by outside specialists. Logistical alliances where manufacturing companies and service providers form distribution networks of various types have been reported from United States (Bowersox 1990) as well as from Europe (van Laarhoven & Sharman 1994). In the transport industry freight forwarders and transportation companies have been able to strengthen their positions by establishing efficient large scale transportation systems as well as local time-table based delivery services. Even in warehousing, activity specialists have increased their importance. Hutt & Speh (1983) report that some large manufacturers have shifted from company-owned (or distributor operated) warehouses to public warehouses. The public warehouse provides economies of scale in operations, as well as specialized services, to meet the particular needs of individual customers. This development within transportation is still going on according to more recent studies (Andersson, 1995 and Johnson & Schneider, 1995). It seems possible, thus, to observe a tendency in transportation similar to what has been identified for information exchange: an increasing out-sourcing of activities and consequently the appearance of transportation brokers.

Low-cost solutions can be offered also by other types of actors. Mail order is one example of a cost efficient channel that can be used when the need for information exchange between buyer and seller is limited. The more experienced the customer is, the less the need will be for exchange of technical information, provided the goods are standardized. The significance of mail-order as an efficient channel for the material flow also for producers goods is that it is used rather extensively in the PC-industry as well as in pharmaceuticals. If mail order is combined with information exchange via Internet then a very cost efficient solution can be attained. The value provided by Internet applications in this respect is analysed by Armstrong & Hagel (1996).

We have found then, that technological development in communication and transportation makes it possible to use intermediaries that are more specialized than industrial distributors have used to be. Traditionally, intermediaries have handled both the flow of information and the flow of goods. Nowadays opportunities are available for activity specialization. In certain applications it seems obvious that such a form of specialization can substitute the prevailing division-of-work between manufacturers and distributors. In cases where the need for exchange of technical information is limited, it is - in principle - possible to bridge the gap between production and consumption by using four activity specialists - one provider of each of the four basic activities: commercial information exchange, administrative information exchange, transportation and warehousing (Figure 2). That would mean a complete restructuring compared to a more traditional distribution structure.

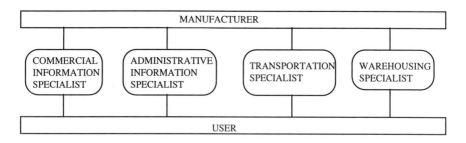

Figure 2. An alternative distribution structure based on activity specialization

It needs to be pointed out, however, that these flows in certain ways are interrelated. Changes in one of the flows will have an impact on the others. It is necessary therefore to take the whole context into consideration when potential changes are analysed. If the context is ignored it might be easy to overlook concurrent changes of importance. Furthermore, the value of the new kinds of resources for exchange of information and goods that now are available is dependent on the way that the resources are combined. As we have seen in Figure 2 specialists have appeared within all the four major distribution activities. This is hardly a coincidence. On the contrary it follows directly from the proliferation of opportunities discussed by Alderson (1954). The opportunity for a firm to specialize in marketing activities depends on the existence of other firms, because the "development of one type of intermediary changes the marketing structure and may prepare the way for still another". The conclusions in our terms are that information specialists is a prerequisite for transportation specialists and vice versa.

Advanced distribution solutions

The low-cost distribution arrangements it will be possible to use for standardized products where end-users' needs of exchange of technical information is limited. Concerning the most advanced distribution solutions manufacturing companies need to establish direct channels to customers. This will make it possible to offer some kind of tailor-made systems. By doing so manufacturers will also come closer to the customer. As a whole, however, it will not be possible for producers in general to be in direct contact with more than a fragmented part of the total end-user base. Therefore, it will be necessary for manufacturers to rely on some form of traditional intermediary even in the foreseeable future. Most business transactions will continue to require more than automated information exchange.

The functioning of these established distribution systems, however, must also be modified. This change is required because manufacturing firms need to get closer to customers. These customers demand more advanced distribution solutions in terms of an augmented product. Therefore manufacturers need prolonged arms that can add value to their offers. The problems perceived by producers in this respect mainly result from cooperation problems between manufacturers and distributors. Traditionally these relationships have been characterized by conflict and confrontation rather than cooperation (Hunt & Ray 1981). The call for more efficient distribution systems will require that such characteristics can be changed. We have earlier identified a number of advantages owing to closer relationships between manufacturer and end-users. It will be important for manufacturers to accomplish these benefits also towards end-users with whom no direct contacts can be established. Long-term cooperative relationships between manufacturers and their distributors therefore should be of great importance to assure the quality of the augmented product and in this way make it possible for manufacturers to get closer to the end-user. A number of authors have emphasized the advantages to manufacturers that can be attained from more integrated producer-distributor relationships (Hardy & Magrath 1988, Narus & Anderson 1987, Rosenbloom 1990). The recommendations for improving performance call for substantial changes in prevailing attitudes and behaviour to make better use of the combined manufacturer-distributor resources.

We will not discuss these changing relationships in general but point to one particular aspect. It seems to be that information exchange is crucial also in these relationships and distributive solutions. One thing that has been a matter of discussion in producer-distributor relationships is the role of distributors as providers of market information. Manufacturers usually perceive industrial distributors as unwilling to share information about their operations and both unwilling and unable to provide information about market areas and individual accounts. According to Webster (1975) the large majority of manufacturers consider their distributors to be of "virtually no help as a source of market information". The reason stated for this ignorance is that distributors are not eager to deliver that kind of information, because they fear to be cut out of business; by the manufacturer either establishing direct distribution to end-users, or switching to another distributor.

But it is fairly obvious that with another kind of relationship climate, based on mutual trust, it should be possible to affect such behaviour. If distributors feel that they are members of a partnership they might behave in another way. This is shown in a case-study which describes a producer-distributor programme that functioned to "cementing this partnership" (Reddy &

Marvin 1986). The programme consisted of two basic parts. One of them was sharing the use of market information, the other one was a programme for manufacturer support services. The case-study clearly shows how both the manufacturer and the distributor will benefit when market information, assembled by the distributor, is used for the further marketing activities undertaken. Such information is an important input into market planning, market segmentation and telemarketing. It is shown in the study that it is possible to affect distributors to share market information with manufacturers if they can see mutual long-term benefits from it. The support services provided by the manufacturer were considered such a benefit that made a distributor "to give up the most sacred of his assets" (Reddy & Marvin 1986).

Another major criticism argued by manufacturers is the lacking distributor competence in technical matters. Even that kind of draw-back can be eliminated, which is shown in an example provided by Cavusgil (1990) in an analysis of distributor training at Caterpillar. When the company was challenged by increasing competition, one of the most important strategic reactions was to make better use of the competence that made the company unique - its very strong dealer network. Caterpillar then decided to make increased efforts to strengthen the total product concept with the services that are important. The quality of these services is totally contingent on the dealers. To utilize the dealer network as a competitive edge, required an extensive support to dealers. Such support takes the form of furnishing sales assistance by providing literature, blueprints, photo's, and so on. Travelling sales engineers can help to demonstrate special equipment. Sales proceedings must be continually updated to reflect changing needs of customers and new applications of products. Dealers are invited to inspect new machines and be briefed about new products. The major theme discussed in the article, however, is the training program introduced. The training program covers technical aspects as well as marketing planning and sales techniques.

It seems evident thus that improved performance in information exchange not always leads to standardization and automation of the processes. On the contrary, in many cases ambitions to make better use of available information might lead firms to develop closer relationships. General benefits of so called information partnerships are discussed in e.g. Konsynski & McFarlan (1989) and Sheombar (1992).

CONCLUSIONS AND IMPLICATIONS

The need for differentiated distribution solutions

Today's dilemma of most manufacturing firms is expressed as being "under increasing pressure to cut costs while at the same time provide even greater customer service" (Oswald & Boulton 1995). We have tried to show that these pressures will call for more differentiated distribution systems. In Figure 3 three very different distribution needs are illustrated, ranging from demand for a low-cost alternative to the most advanced solution.

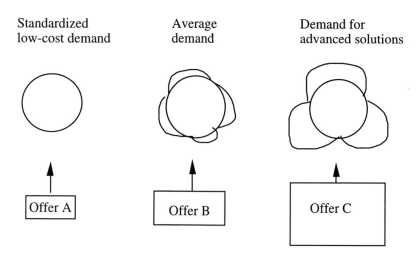

Figure 3. Differentiation of distribution offers in accordance with customer needs

A producer aiming at satisfying each of these demands must develop three different distribution concepts. If only offer B is used it will not be possible to satisfy either the demand for low-cost distribution or the need for the more advanced solutions.

The problem is then to identify the most appropriate distribution solution for each type of distribution need. The availability of new channel alternatives provides opportunities for producers to design customer adapted solutions. In many cases, however, implementation of such changes has been difficult to undertake. Distribution systems in general are characterized by stability owing to established industry structures and historical patterns. However, in some industries major changes have been undertaken. One example is the computer industry. As late as 1981 IBM used virtually no indirect channels. In 1985, however, IBM was using a variety of intermediaries such as "distributors, retail computer dealers, «value-added dealers«, and «complementary marketing organizations«" (Cespedes 1988). These channels accounted for an increasing proportion of IBM's revenues and the majority of revenues for products like personal computers, smaller business systems, soft-ware and various kinds of supplies. Some distributors also had become important channels for higher unit-priced IBM-products (minicomputers, network installations and even parts of the main-frame computer line). These things have developed even more over time and given rise to very complex distribution networks in the computer industry. Firms are engaged in numerous strategic alliances and partnerships in different constellations. The borders between producers and distributors are becoming unclear. Producers are nowadays working in direct contact with end-users that were earlier served by distributors. Producers might utilize other manufacturers' hard-ware in system solutions of various kinds. Distributors supply complex systems of hardware, soft-ware and services in a role as system integrators. The magnitude of these changes is very substantial (see for instance Tunisini 1994).

It has also been reported that Japanese distribution systems are under-going structural changes. Distribution systems in Japan have been perceived as rather closed and a reason for the difficulties for foreign manufacturers in most industries to enter the Japanese market.

Recent economic pressure however has forced Japanese manufacturers to "lock up" their distribution systems. The most important reason for the changes has been to give way to new low-cost channels. Means of achieving these goals have been "bypassing existing distribution systems" and "establishing relationships with outside distributors" (Shill et al. 1995).

When analyzing changes in distribution we have emphasized the need to use a combined perspective including the strategic ambitions of manufacturers, the opportunities provided by IT and the requirements from customers. It is indeed impossible to separate them from each other to understand the on-going as well as the potential changes. Nevertheless, in a chapter on IT and distribution we feel it necessary to say something more about the role of IT per se. So far we have mainly discussed IT from the perspective of manufacturers. Now we will change the view and consider the effects on a more aggregate level.

IT and distribution structures

We have shown a number of significant improvements of distribution performance owing to the implementation of various IT-solutions. However, the effects seem to have been of less magnitude than was expected ten years ago. So far information technology has not contributed to major restructuring of distribution. One reason for the rather low IT-impact is that "information technologies have been used for over three decades primarily to automate age-old business procedures" (Teng et al. 1994). According to that study the IT-systems available today can be deployed to fundamentally change business processes. This is the same as to say what was earlier stated that IT will have an impact when it can support and promote other strategic issues of a company. It is in the inter-play between these strategic isssues and the available IT-tools that effects will be created. Similar opinions have been expressed elsewhere (Sheombar 1992):

> Instead of using EDI for the automation of the current information flows between organizations, one should first question whether the current way of coordinating should be prolonged.

That author therefore asks for new distributive arrangements rather than automation of the present ones. To be able to say something about potential re-organization of distribution structures we need some kind of framework to discuss from. Elsewhere we have argued for a network approach to distribution analysis (Dubois et al. 1996). The basic building blocks of an industrial network are actors, activities and resources.

Information technology basically is a new resource made available for the functioning of a distribution network. The major thing with this new resource is that it makes it possible to improve the activity structure of the network. The case of Atlas Copco Tools is a significant illustration of improved efficiency due to changes of activity structures when implementing centralized warehousing. The change towards centralized warehousing restructured distribution in a rather fundamental way by rearranging the activity structure for creation of "availability" in distribution. For a manufacturer to be competitive, it is necessary that products are available when customers demand them - otherwise that potential order might be lost to a competitor. Traditionally, inventories close to the market have been the most important means to provide availability. The major reason is that the planning capability of customers has been limited, at the same time as the lead time from order placement to delivery has been substantial. Inflexible manufacturing operations and information systems with limited capacity have been the major draw-backs. The characteristics of manufacturing systems also made it necessary to produce large series to obtain economies of scale. Also that kind of behaviour resulted in huge inventories. The total investments in inventories, therefore, were considerable. Sometimes, stocks could be held by firms on three or four levels in the distribution system. This was a very expensive way of assuring availability (Figure 4).

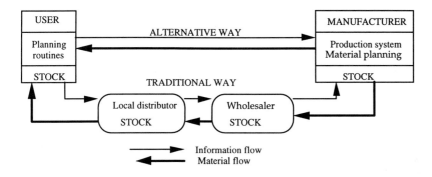

Figure 4. Two different activity structures for creation of availability (Gadde 1994:246)

During the 1980s, reducing capital costs was given priority in many firms. To achieve that objective, alternative activity structures to create availability were required. More flexible manufacturing systems were developed with the basic aim to decrease the need to produce for stocks to obtain economies of scale. In combination with increasing efficiency in handling internal material flows, the new manufacturing techniques made it possible to reduce lead times in a substantial way. These factors, together with developments in information technology and more efficient transportation systems, have formed a new alternative to assure availability. In theoretical terms, thus, speculation has been replaced by postponement. As has been shown earlier, inventories can be decreased in a substantial way without sacrificing customer service. The effects of this kind of time-based competition are explored in Abrahamsson (1993).

Such changes in the activity structure will affect the importance of the various activities. Our example has shown that information handling and transportation have become more important than before and stock-holding less important. The changes undertaken will also affect the need for coordination of the activities when buffers are more or less eliminated.

The changes of the activity structure will have an impact also on the value of the resources in the network. Inventories have been considered a very important resource because inventories guaranteed availability. The value of this resource is diminishing when other ways to secure availability are established. On the other hand joint information systems, efficient transportation vehicles, sophisticated warehousing equipment and competence and capability of people will become more important resources in the type of activity structures that now are created. The value of a specific resource thus will be contingent on the way it is combined with other resources.

Resources are controlled by actors. Consequently, changes in the value of different resources will affect the actor structure. Compared with manufacturers, intermediaries have had a more adequate resource-base for serving small and medium-sized customers. Technological development and the entrance of new actors might change these conditions and make new structures emerge. Furthermore, stock-holding has been an important activity for traditional distributors. If this activity is decreasing in importance, the position of intermediaries will be affected. Moreover, in a number of industries distributors' profits partly have relied on income from a systematic utilization of price fluctuations in stock-holding operations. In this way the on-going changes might erode the resource base of present intermediaries, as speculative inventories will have much less place in the alternative type of activity structure. It has even been argued that the existence of speculative inventories is the fundamental reason for title-holding institutions between production and consumption. One of the most prominent researchers within distribution states that "without such inventories there may be little economic justification for a title holding intermediary to enter the channel" (Bucklin 1965).

The actor structure will be affected also concerning the relationship between the actors. As was briefly discussed previously, manufacturers probably will be interested in developing more collaborative relationships with distributors. Such relationships will be a prerequisite for assuring the quality of the augmented product and make it possible for producers to get closer to those customers that will continue to be served by intermediaries.

Final remarks

Distribution systems for industrial goods are under major re-organization. The magnitude of the potential changes can be substantial, owing partly to the strong pressure for improvements and partly to the opportunities made available through technological development. In this process the development of IT can play a significant role because it can improve the efficiency of the information flow as well as the coordination of the material flow. In this way IT can make substantial contributions to re-organization of business processes.

So far the effects of IT have been of less importance than was expected. Some authors have even argued that the understanding of the effects of EDI on organizations as late as in 1992 was "still at a rudimentary level" (Holland et al. 1992). But maybe it is misleading to conclude that the effects have been less than they should have been. Maybe the expectations were unrealistic? One author states that the "history of IT can be characterized as the overestimation of what can be accomplished immediately" (Strassman 1985). Bearing in mind that distribution systems are difficult to change owing to dependencies among firms, one should not be surprised that changes take time. It is interesting to note also that the most significant effects of changing distribution structures come from the computer industry. This industry had less historical patterns to rely on. Therefore it was easier to introduce alternative structures than it has been in more established and "cemented" distribution networks.

The analysis has shown that IT is an important resource that can contribute to rationalization of distribution. IT makes it possible to affect the efficiency of activity structures. We have tried in another article to analyse the conditions for and the potential effects of changes in activity structures in distribution networks (Dubois et al. 1996). According to that analysis the potential for change is very situation-specific and dependent on the characteristics and resources of the counter-parts in the network. It is not possible therefore to say something about changes in general. As a whole, however, a number of illustrations point to the fact that substantial improvements have been possible to attain. We would agree therefore not only with the first conclusion in Strassman (1985) regarding the short-term over-estimation of IT-impact, but also with the second one presented. According to Strassman the second experience from the history of IT-impact is an "underestimation of long-term consequences".

These long-term consequences generally are difficult to predict. They will only result from the interaction when firms start to try to affect the activity structures. If it is difficult to say something about potential changes in the activity structure in general, it will be even more difficult to predict changes of the actor structure. However, we have identified opportunities for manufacturers to improve performance of their distribution operations. It is likely that such rationalization efforts partly will take place through offensive use of new channel alternatives. In this process the traditional form of intermediary - an independent actor covering all distribution functions - seems to be challenged in a number of ways. They will be challenged by specialized firms who can be more effective in taking care either of the flow of information or the flow of goods. They will be challenged also by manufacturers who will favour "prolonged arms" rather than independent distributors. However, this will not necessarily mean a threat to existing intermediaries in a distribution system. On the contrary they can make use of the new opportunities and in this way strengthen their positions. All depends on how they deal with the new conditions.

In a study of the automobile distribution system the analysis points to the need for a radical restructuring of distribution (Mercer 1994). The author arrives at the conclusion that unbundling of current dealer activities (new car sales, used car sales, services, parts sales) is a prerequisite for performance improvements. The major reason being that these four business areas of a dealer are very different in terms of scale economies and resources and capabilities required. Franchised dealers have been challenged by specialized firms in each of these areas. A development towards unbundling would be in line with what we have identified as a specialization in the activity structure. Obviously such a change would mean problems for the established dealers.

In the pharmaceutical industry, on the other hand, it seems as if the same forces (reducing costs and improving services) have led to a strengthened position for intermediaries (Oswald & Boulton 1995). According to these authors drug distribution can serve as a show-case illustrating how wholesalers can improve their network positions. The basis for building industry position has been an emphasis on "mega distribution centers, new information technologies, rapid response capabilities, voluntary retail programmes, mail-order and target marketing". Large-scaled mega distribution centers have been established as means for achieving economies of scale. EDI is used by about half of the firms in the industry for sending and receiving administrative information. McKesson Drugs still seems to lead the industry in information management operations. According to the authors they have created a nation-wide integrated care information system. To do this McKesson has "merged its prescription information network with medical and pharmaceutical outcome research" (Oswald & Boulton 1995).

The final conclusion to be drawn is that IT provides opportunities for restructuring of distribution. Our examples have shown that these opportunities have been taken care of by firms on different levels in traditional distribution systems as well as by firms outside. It seems as if the on-going changes might lead to major structural effects making traditional categorization of firms obsolete. Such a conclusion is further strengthened by the findings that the major impact on efficiency has been attained when the border between what usually was called production activities and distribution activities has been changed.

REFERENCES

Abrahamsson, M. (1993). Time-Based Distribution. The International Journal of Logistics Management, Vol. 4, No. 2, pp. 75-83.

Alderson, W. (1954). Factors Governing the Development of Marketing Channels. In Clewett, R. (ed) Marketing Channels for Manufactured Products. Richard D. Irwin, Homewood, Ill.

Alexander, R., Surface, F. and Alderson, W. (1940) Marketing. Ginn and Company, Boston.

Andersson, D. (1995). Logistic Alliances and Structural Change. Linkšping Studies in Science and Technology, No. 470. Linköping University, Department of Management and Economics.

Armstrong, A. & Hagel, J. (1996). The Real Value of On-Line Communication. Harvard Business Review, May-June, pp. 134-141.

Bagchi, P. (1992) International Logistics Information Systems. International Journal of Physical Distribution and Materials Management, Vol.22, No.9, pp. 11-19.

Bardi, E. & Tracey, M. (1991). Transportation Outsourcing. A survey of US practices. International Journal of Physical Distribution and Logistics Management, 21,3, pp. 15-21.

Bowersox, D. (1990). The Benefits of Strategic Alliances. Harvard Business Review, July-August, pp. 36-43.

Bruce, L. (1987). The Bright New Worlds of Benetton. International Management, November, pp. 24-35.

Bucklin, L. (1965). Postponement, Speculation and the Structure of Distribution Channels. Journal of Marketing Research, vol. 2, pp. 26-31.

Carlisle, J. & Parker, R. (1989). Beyond negotiation. Redeeming customer supplier relationships. John Wiley & Sons, Chichester.

Cavusgil, T. (1990). The importance of Distributor Training at Caterpillar. Industrial Marketing Management, vol. 19, pp. 1-9.

Cespedes, F. (1988). Channel Management is General Management. California Management Review, Fall, pp. 98-120.

Corey, R. (1985). The Role of Information and Communication Technology in Industrial Distribution. In Buzzel, R (ed) Marketing in an Electronic Age. Harvard Business School Press, Boston.

Corey, R., Cespedes, F. and Rangan, K. (1989). Going to Market. Distribution Systems for Industrial Products. Harvard Business School Press, Boston.

Daugherty, P. & Dröge, C. (1991). Organisational Structure in Divisionalised Manufacturers: The potential for outsourcing logistical services. International Journal of Physical Distribution and Logistics Management, vol. 21, No. 3, pp. 22-29.

Dubois, A., Gadde, L-E & Mattsson, L-G (1996). Activity Structures in Distribution - a framework for analyzing efficiency. In Ghauri, P. (ed) Advances in International Marketing. JAI Press, Cambridge.

Fletcher, K. (1992). Database Marketing in the UK Automotive Industry: Some empirical evidence. Proceedings of the 21st annual meeting of EMAC. •rhus university.

Gadde, L-E & HŒkansson, H. (1993.) Professional Purchasing. Routledge, London.

Gadde, L-E & Mattsson, L-G (1987). Stability and Change in Network Relationships. International Journal of Research in Marketing, No. 4, pp. 229-41

Gadde, L-E (1994). Developments in Distribution Channels for Industrial Goods. In Baker, M. (ed) Perspectives on Marketing Management. John Wiley & Sons, Chichester.

Hardy, K. & Magrath, A. (1988). Ten Ways for Manufacturers to Improve Distributor Management. Business Horizons, November-December, pp. 65-69.

Harrington, L. & Reed, G. (1996). Electronic Commerce (finally) Comes of Age. The McKinsey Quarterly, No. 2, pp 68-77.

Holland, C., Lockett, G. and Blackman, I. (1992). Planning for Electronic Data Interchange. Strategic Management Journal, Vol. 13, pp. 539-550.

Hunt, S. & Ray, N. (1981). Behavioural Dimensions of Channels of Distribution : The state of the art. Proceedings of the 8th international research seminar in marketing. IAE, Aix-en-Provence.

Hutt, M. & Speh, T. (1983). Realigning Industrial Marketing Channels. Industrial Marketing Management, vol. 12, pp.171-77.

HŒkansson, H. (ed) (1982). International Marketing and Purchasing of Industrial Goods. John Wiley, Chichester.

Johnson, J. & Schneider, K. (1995). Outsourcing in Distribution: The Growing Importance of Transportation Brokers. Business Horizons, November-December, pp. 40-48.

Johnston, H. & Vitale, M. (1988). Creating Advantage with Interorganizational Information Systems. MIS Quarterly, Vol. 12, No. 2, pp. 153-166.

Kaufman, F. (1966). Data Systems that Cross Organizational Boarders. Harvard Business Review, January-February, pp. 141-55.

Konsynski, B. & McFarlan, W. (1980). Information Partnerships - Shared Data, Shared Sales. Harvard Business Review, September-October, pp.114-120.

Lamming, R. (1993). Beyond Partnerships. Prentice Hall, Hemel Hempstead, UK.

van Laarhoven, P. & Sharma, G. (1994). Logistic Alliances: The European Perspective. The McKinsey Quarterly, No. 1, pp. 39-49.

Mallen, B. (1973). Functional Spin-Off: A Key to Anticipating Change in Distribution Structures. Journal of Marketing, vol. 37, pp. 18-25.

Malone, T., Yates, J. and Benjamin, R. (1989). The Logics of Electronic Markets. Harvard Business Review, May-June, pp. 98-103.

McKenna, R. (1988). Marketing in an Age of Diversity. Harvard Business Review, September-October, pp. 88-95.

Mentzer, J. & Gandhi, N. (1993). Expert Systems in Industrial Marketing. Industrial Marketing Management, Vol. 22, pp. 109-116.

Mercer, G. (1994). Dont just Optimize - Unbundle. The McKinsey Quarterly, No. 3, pp. 103-116.

Morgan, J. (1991). Competition Comes to the Distributor. Purchasing, May 16, pp. 64-71.

Moriarty, R. & Swartz, G. (1989). Automation to Boost Sales and Marketing. Harvard Business Review, January-February, pp. 100-108.

Narus, J. & Anderson, J. (1987). Distributor Contributions to Partnerships with Manufacturers. Business Horizons, September-October, pp. 34-42.

Nieschlag, R. (1954.) Die Dynamik der Betriebsformen in Handel. Schriftenreihe Neue Erfolge Nr. 7. Rheinish - WestfŠliges Institut fur Wirtschaftsforschung, Essen.

Oswald, S. & Boulton, W. (1995). Obtaining Industry Control: The Case of the Pharmaceutical Industry. California Mangement Review, Vol. 38, No. 1, pp 138-162.

Porter, M. & Millar, V. (1985). How Information Gives You Competitive Advantage. Harvard Business Review, July-August, pp. 149-160.

Reddy, N. & Marvin, P.(1986). Developing a Manufacturer - Distributor Information Partnership. Industrial Marketing Management, vol. 15, pp. 157-163.

Rosenbloom, B. (1990). Motivating Your International Business Partners. Business Horizons, March-April, 53-57.

Scott-Morton, M. (ed) (1991). The Corporation of the 1990s - Information Technology and Organizational Transformation. Oxford University Press, London.

Sheombar, H. (1992). EDI-induced Redesign of Co-ordination in Logistics. International Journal of Physical Distribution and Logistics, Vol. 22, No. 8, pp. 4-14.

Shill, W., Guild, T. and Yamaguchi, Y. (1995). Cracking Japanese Markets. The McKinsey Quarterly, No. 3, pp. 32-40.

Spar, D. & Bussgang, J. (1996). Ruling The Net. Harvard Business Review, May-June, pp. 125-133.

Stern, L. & Sturdivant, F. (1987).Customer Driven Distribution Systems. Harvard Business Review, July-August, pp. 34-41.

Strassman, P. (1985. Information Payoff. The Transformation of Work in the Electronic Age. The Free Press, New York.

Teng, J., Grover, V. and Fiedler, K. (1994). Re-designing Business Processes Using Information Technology. Long Range Planning, Vol. 27, No. 1, pp. 95-106.

Tunisini, A-L (1994). Developing an Effective Distribution Channel in the Computer Industry. In Bierman, W. and Ghauri, P. (eds) Meeting the Challenges of New Frontiers, pp. 592-611. Proccedings of the 10th IMP Annual Conference. University of Groningen.

Verity, J. (1994). A Potent New Tool for Selling: Database Marketing. Business Week, September 5, pp. 34-40.

Voorhees, R. & Coppett, J. (1983). Telemarketing in Distribution Channels. Industrial Marketing Management, vol. 12, pp. 105-112.

Webster, F.(1975). Perceptions of the Industrial Distributor. Industrial Marketing Management, vol. 4, pp. 275-284.

4

INTERORGANIZATIONAL LOGISTICS FLEXIBILITY IN MARKETING CHANNELS

Arne Jensen

ABSTRACT

In recent years, considerable attention in channel research has been devoted to the timing of logistics activities in marketing channels in the form of JIT management and time-based competition, i.e. phenomena that are characterized by temporal rigidity. On the other hand, research on its opposite - logistics flexibility - has been conspicuous by its absence, even though logistics flexibility may well represent a potential source of improved efficiency.

This paper develops and analyzes the concept of interorganizational logistics flexibility (ILF) in marketing channels and examines empirically the extent to which manufacturing and wholesale companies in channels can offer carriers a temporal flexibility that can be used to increase resource utilization and, thus, efficiency in the channels. Finally, the required information systems and other aspects of implementation are discussed.

Keywords: Flexibility, logistics flexibility, interorganizational logistics flexibility, flexibility in marketing channels, resource utilization of line-based road carriers, flexible deliveries, flexible pickup, interactive booking, flexibility information in road haulage, flexibility in freight demand

INTRODUCTION

Actors involved in marketing channels are increasingly trying to reduce uncertainty in their operative activities by demanding quick responses that are firmly fixed in time, or at least predictable, by other actors in the channel. In logistics channels, the way of thinking has of later years been influenced particularly by the JIT philosophy and ideas of time-based competition, which has led to more stringent control of activities in time and space (see Frazier et al., 1988; Billesbach & Hayen, 1994; Daugherty & Pittman, 1995; Hise, 1995; Stalk & Hout, 1990, Carter et al., 1995). These demands, which have come principally from the primary actors in the channels, i.e. manufacturers, wholesalers, and retailers, have brought on a trend toward increased temporal rigidity in the performance of operative logistics activities.

In certain markets, the channels' specialized suppliers of logistics services, such as carriers, have responded to this trend by improving quality in the temporal dimension and by standardizing their services, such as rapid, standardized shipments. A combination of customer demand, the competitive situation, and scale advantages resulting from standardization are the presumable reasons why, in many cases, suppliers are giving their customers a level of logistics service that exceeds need. The consequence of this has been unnecessary utilization of resources, seen from the overall channel perspective. Thus, paradoxical as it may seem, it is probable that despite stricter time management a temporal flexibility in interorganizational relations in marketing channels exists or can be created, and it may be expected to hold the potential for increased efficiency throughout the channel. Thus, it should be possible to take advantage of this flexibility without actually affecting the level of logistics customer service in the channels.

This background description indicates a need for both positive and normative scientific knowledge about flexibility. The literature on flexibility in marketing channels is very limited, however, and more research is needed. A suitable research strategy would be (1) to identify flexibility conceptually and empirically and to ascertain its extent and character, (2) to estimate the costs and benefits related to the utilization of flexibility, and (3) to study the design and behavioural aspects of systems for flexibility information and flexibility management.

This paper seeks to contribute to point 1 above by examining the extent to which manufacturers and wholesalers in their functions as shippers in marketing channels can offer carriers temporal flexibility that can help to increase channel efficiency. Finally, the paper discusses information systems that can facilitate the utilization of shippers' flexibility in marketing channels.

THE TERM "FLEXIBILITY" IN THE LITERATURE

Research on flexibility has been carried on within several disciplines, such as economics, business administration, and ecology. Of these, research on business administration is of the greatest interest for this paper.

The literature on marketing channels seems to use the term "flexibility" only as a general descriptive adjective. One exception, however, is Andersson (1992) who refers, among other things, to an article by Gadde & Håkansson (1992) on change and stability in marketing channels. Andersson seems to use the term "flexibility" primarily as a synonym of change and advocates using the theory of loosely coupled systems (Weick, 1976) when studying stability and flexibility in channel structures. However, his article lacks a deeper penetration of the flexibility concept and its field of application is outside the scope of this study. The type of flexibility involved in the present study does not seem to have been dealt with in the literature on channels.

Research on strategic flexibility has been carried on by Ansoff (1975), Eppink (1978), Krijnen (1979), Harrigan (1980 and 1985), Evans (1991), Das & Elango (1995), and Aaker & Mascarenko (1984, p. 74). Aaker & Mascarenko define strategic flexibility as:

"... the ability of the organization to adapt to substantial, uncertain, and fast occurring (relative to the required reaction time) environmental changes that have a meaningful impact on the organization's performance"

and they mention three ways to achieve strategic flexibility: diversifying, investing in underused resources, and reducing commitment of resources to a specialized use (pp. 74-82). This definition provides a representative picture of research on strategic flexibility. It deals with the way companies create flexibility that facilitates their own responses to strategic surprises or rapid changes, while the focus of the present study is on operative flexibility that channel members can offer other channel members in order to increase total channel efficiency when the latter carry out well-defined distribution activities. In this way, the role of flexibility is not only to absorb uncertainty, as in strategic flexibility, but also to proactively reduce known variations. Consequently, the research on strategic flexibility is of limited interest in the present study.

Research on flexible manufacturing systems usually examines flexibility as an adaptive response to uncertainty in the environment (Gupta & Goya, 1989), but it can also be used in proactive strategies, according to Gerwin (1993). He identifies five types of manufacturing flexibility that seem to be typical of this research: production mix flexibility as a buffer against the uncertain distribution of demand within a line of manufactured goods, change over flexibility for handling uncertainty concerning the life cycle of a product, modification flexibility as a guard against uncertainty over the need to adapt new products to the customer, volume flexibility for countering uncertainty over the magnitude of demand, rerouting flexibility to compensate for machine downtime, and material flexibility for dealing with uncertainty over the availability of materials. In addition to the authors mentioned, Hutchinson (1973), Buzacott (1982), Adler (1988), De Meyer et al. (1989), Hall & Tonkin (1990) and others have dealt with flexible manufacturing. Research on manufacturing flexibility is also of limited relevance to the present study except for the possible interpretation that flexible manufacturing in the future may help creating logistics flexibility for other actors.

The concept of flexibility has also aroused interest in decision theory research (Merkhofer, 1977, Pye, 1978, Mandelbaum and Buzacott, 1990, and others) regarding the value of flexibility in decisions under uncertainty in sequential decision problems, for example. In this research, flexibility is usually defined in terms of the number of alternatives to choose from at a later date once an initial decision has been made. The value is analyzed theoretically using formal mathematical decision models. Decision theory research on flexibility is also of extremely limited value to the present study.

CONCEPTUAL FRAMEWORK

Marketing channels are seen here as sets of interdependent organizations involved in the process of making a product or service available for use or consumption (Stern et al., 1996, p. 1). This is a systems view that combines the two approaches that Gattorna (1978) called the institutional and the functional. We have actors (organizations) who carry out distribution functions (marketing flows). Each marketing flow consists of activities that are performed using resources controlled by the actors. This view from channel theory also coincides with the model of industrial networks presented by Håkansson & Johansson (1992, p. 29). They explicitly distinguish subsystems (networks) of actors, activities, and resources. The focus in the present study is on logistics activities in a marketing channel. Thus, for descriptive reasons, it has been found useful to distinguish two subsystems in a channel, a logistics channel, where the key logistics activities of transportation, inventory management, and handling take place, and a transaction channel engaged in negotiation, contracting, and post transaction administration of sales. The reasoning behind this approach - structural separation - is that the logistics channel and the transaction channel can, in part, consist of different actors (see Bowersox et al., p. 198).

Let us now define the key concept of this paper, interorganizational logistics flexibility (ILF).

Interorganizational logistics flexibility (ILF) refers to key logistics activities specified and ordered by one set of actors in the channel and executed by another set of actors belonging to the logistics channel. The degree of flexibility is determined by the size of the choice set for execution and planning in primary logistics dimensions that is available for the executive set of actors.

In this definition, of course, "a set of actors" may consist of only one actor. The terms *executive flexibility* and *planning flexibility* will be used, depending on whether the flexibility refers to the execution or the planning of activities. The primary logistical dimensions are time and quantity, but others may also appear, such as the physical conditions (temperature, vibrations, etc.) under which the logistics flow is executed, possible packaging and load carriers that are acceptable, etc.

An individual logistics activity includes a certain flow of goods in the channel and it takes time to execute, which can be described by the time distribution of the quantity of goods. Thus, in the primary dimensions, the degree of executive flexibility increases with the number of different time distributions of the given quantity among which the executive actors may choose.

Planning flexibility increases in the primary dimensions with increasing possibilities for advance planning and with the number of activities that can be planned in advance. Below, we will deal only with interorganizational logistics flexibility in the primary dimensions of time and quantity, which is called *temporal flexibility*, since time is the driving factor, and quantity distribution is often dependent on the temporal factor.

Of course, interorganizational flexibility is a relevant concept in practice because increased flexibility can be expected to create a potential for lower costs, increased efficiency, or improved output quality, measured at a suitable point in the logistics flow. The relationship between temporal flexibility and resource utilization is of particular interest. Resource needs for a certain logistics activity depend on short-term variations in demand for the activity in question and they are proportional to the maximum demand. This applies to the resources fixed facilities, vehicles, equipment, and personnel. Increased executive flexibility can be used for demand smoothing, which reduces maximum demand and, thus, resource needs. Increased planning flexibility gives the executive actors a greater chance to reallocate resources among the various activities and to reduce resource-demanding uncertainty.

THE EMPIRICAL ARENA

Resource utilization by line-based road carriers

For a long time and to an increasing extent, the primary actors in Swedish marketing channels have left the execution of transport-related activities to specialized carrier companies. This example of what Mallen (1973) refers to more generally as "functional spin-off" means, in this case, that manufacturers and wholesalers with medium and high quality goods generally use line-based trucking. To a limited extent, road carriers are also taking over other logistics activities. Their cooperation with shippers is based on contractual agreements, and these carriers are important members of the marketing channels in which they operate.

Line-based road traffic is organized and marketed by just a few major surface forwarders, each with its own network of lines and its own terminals. Pickup and distribution traffic around the terminals as well as long-haul traffic between terminals are accomplished by truck. Rail is used to a limited extent in long-haul traffic, mostly in the form of piggyback traffic. The entire Swedish market has access to line-based road transport. A surface forwarder has a number of motor carriers under contract, and the forwarder and its carriers may be regarded as one enterprise. Each motor carrier is operatively responsible for traffic over one or a few lines. The level of service is high and standardized. Transit times are such that a shipment leaving the sender on day 1 will arrive at the recipient on day 2 at distances shorter than ca 700 km, i.e. for a very high percentage of shipments.

It has been claimed that the domestic line-based trucking operates at a low level of resource utilization. Studies on load capacity utilization also indicate that this is the case. For example, Jensen (1990, ch. 4) examined the flow of goods over 22 major long-haul lines in Sweden, with line distances greater than 200 kilometers. The results of this study indicate that the load factor on these lines, measured in freightage weight, in 1983 was 60 to 70%. This and other studies indicate that, at least theoretically, the potential for increase in load capacity utilization is at least 30%.

The low utilization of resources in domestic line-based road transport is due to the complex interaction of variations in demand, late booking, the competitive situation, and traffic organization. Demand from shippers varies systematically with the economic situation, the season, and the day, and there are significant random daily fluctuations. Variations from day to day are greater than those related to the economy and to the season. There are also imbalances, i.e. the flow of goods is not equal in both directions over a line. The quantity booked often differs from that which is actually utilized. Under all these demand conditions, a carrier who wishes to avoid lost demand is forced to have an available capacity that far exceeds the average transport needs, since his capacity must be designed for the maximum load. Consequently, capacity in the form of vehicles, personnel, handling equipment, and terminals must be over-dimensioned. This results in low resource utilization.

Strategies for increasing resource utilization

Theoretically a carrier could increase resource utilization on a line by decreasing capacity and giving up orders that temporarily exceed capacity. However, this strategy is prevented by the competitive pressure. The carrier on line A-B of forwarder X is unwilling to give up an order, only to have it shipped by route A-C-B by another carrier within X and, in the competitive situation that exists (oligopoly), forwarder X will not accept giving up an order, if the result would be that the order is lost to another forwarding company. The existing market and organizational structure seems to block strategies that would increase resource utilization if such strategies also involve a risk of shifts in market shares for carriers or forwarders.

Another strategic approach that need not result in altered market shares would be to try to utilize the temporal flexibility of the demand for transportation. Given that such flexibility exists and that it is sufficient, it should be possible to use it to even out load variations or to increase advance planning, which can create the proper conditions for increased resource utilization. Jensen's study (1990) indicates that there is a certain amount of flexibility in the demand for transportation. He found, first of all, that shippers on 12 key lines in Sweden overestimated transport times - the actual average transport time was shorter than they estimated - and secondly that the estimated transport time, in turn, was shorter on average than what the shippers considered to be the longest acceptable time. It is believed that the reason for this is that freight companies have too little information concerning their customers' transportation quality needs. This, combined with the oligopolistic competitive situation and demands from a small group of powerful, vocal shippers, has forced a level of quality that, on average, exceeds customer needs. This paper will examine the possibilities of the flexibility approach in greater detail.

The flexibility strategy: Impact and concept operationalization

As mentioned, the goal of freight carriers in line-based traffic is to satisfy all demands themselves. Figure 1 shows a model of demand concepts that, under these circumstances, determine the transport capacity that the carrier must have available for a line. The daily demand curve in the figure is assumed to represent the traffic direction with the maximum demand. In this traffic direction, each day of departure the carrier must have a capacity V_2 that is sufficient to cover the greatest conceivable demand. In addition, there is a margin of uncertainty, $V_3 - V_2$, since the maximum daily demand cannot be predicted with certainty when the transport capacity is decided. The carrier's total required capacity on the line is proportional to V_3.

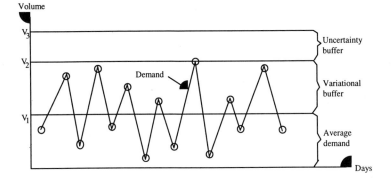

Figure 1. Demand concepts as determinants of capacity on a freight line

Meaningful flexibility concepts are those that represent possibilities for reducing the variational buffer and the uncertainty buffer. The variational buffer can be reduced by demand smoothing, which would be possible if pickup from the sender and delivery to the recipient could be determined within time intervals, rather than at points of time, possibly in combination with a time interval for the transit time. The uncertainty buffer can be reduced by having shippers supply carriers with forecast information or book transport earlier, which could help to reduce the uncertainty buffer by providing greater lead times for moving vehicles, handling equipment, and personnel between lines and tasks and by facilitating planning. Theoretically, flexibility strategies could produce a maximum reduction in capacity needs to a level equal to the average demand, V_1 in Figure 1.

In this light, the following flexibility concepts were found to be of interest in the empirical study:

Planning flexibility:

- earlier booking for shipments;

- earlier forecast information on transport needs.

Executive flexibility:

- flexible pickup times for the sender;

- flexible delivery times for the recipient;

- longer transit times for the goods.

These five types of flexibility can all help to create the conditions for reducing load variations and for earlier planning and, thus, increased resource utilization and greater cost efficiency, without any reduction in the level of logistic customer service in the channel. The key concepts in the empirical study are illustrated in Figure 2.

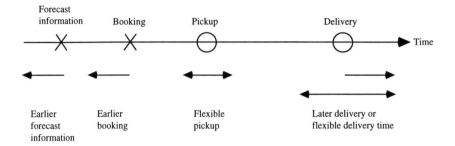

Figure 2. Flexibility concepts in the empirical study

RESEARCH DESIGN

A questionnaire was designed, after a preliminary study, which included in-depth interviews with transport managers with three manufacturers and four wholesalers. The questionnaire contained operationalizations of flexibility concepts, background variables, and explanative variables. The purpose was to measure the extent to which it is possible to increase the flexibility of transport activities with respect to the five defined flexibility concepts, to determine attitudes toward increased flexibility and what incentives would be required, and to seek limited explanations as to the relationships that affect flexibility.

The questionnaire was tested thoroughly at the companies included in the preliminary study and then sent out to a stratified random sample of 1,381 companies, stratified as manufacturing companies and wholesalers and by company size, making a total of 2x9=18 strata. The population comprises all manufacturing and wholesale companies in Sweden. The term "company" used here is materialized by the sampling element "local unit" used in official statistics, which is also relevant from a logistical standpoint. When the questionnaire was sent out, it was not known whether the selected companies used domestic trucking lines.

After one reminder, the response frequency was 45%, 25% of which did not use domestic trucking lines. These were excluded from the sample, since they did not belong to the target population.

A nonresponse analysis was performed among 212 randomly selected companies that failed to respond after one reminder. These were sent a second small questionnaire asking why they had failed to respond. The companies that also failed to respond to this questionnaire were contacted by telephone. This process resulted in a total response rate of 90% in the nonresponse analysis. Among these companies, 35% indicated that they did not use transport by domestic trucking. 15% of the companies did not respond because the company had been dissolved, bought, or made only limited use of transport services. Thus, these could be excluded from the sample. 30% of the companies indicated that they used transport by domestic trucking, but that they did not wish to respond to the questionnaire due to lack of time.

The reasons and explanations that were found in the nonresponse analysis gave no cause to assume that those who did not answer differed in any appreciable way from those who responded. The structure of the nonrespondents with regard to number of employees per company gives us the same impression. Moreover, a significance test was made. The companies that were selected in the nonresponse analysis were called on to answer the original questionnaire, which resulted in a sub-sample of 23 completed questionnaires. The means of 24 important variables from this sub-sample were tested for significance against the corresponding means for those companies that responded at or before the time of the first reminder. Statistically significant differences were found for only two variables, and no systematic patterns could be observed.

All in all, these arguments indicate that the nonresponse sector of the random sample did not differ from the responding sector and that the respondents can be seen as a stratified, random sample of the entire population, including the part of the population that would not have responded to a hypothetical complete census. Based on the nonresponse analysis, the 23 questionnaires from the sub-sample were combined with those from the initial respondents to form a stratified sample of 296 companies. The results of the study were estimated from this sample.

The population's stratum sizes are used as weights in estimating the stratified mean values. These have been estimated for each stratum as the number of companies in the population multiplied by the proportion of the random sample that makes use of freight lines. It should be observed from here on that all mean values and proportions included in the results of the study represent the population of manufacturers and wholesalers *who utilize freight lines.*

RESULTS OF THE SURVEY

Unless otherwise stated, the results apply to the entire random sample. In certain cases, results are divided into wholesalers and manufacturers or given for large (over 50 employees), medium-sized (5-49 employees), and small (0-4 employees) companies. The sample's distribution according to these categories is shown in Table 1.

Table 1. Distribution of sample by company size and company category

Company size	Company category	
	Wholesaler	Manufacturer
Small	35	25
Medium	34	28
Large	43	131

When subgroups have been found different at a statistical significance level of $p=0.06$, the difference will be referred to as significant.

The booking process

In the Swedish trucking line business, shipments that exceed certain weight and volume limits are classified as full loads. In this study, other shipments are classified as partial loads. Full loads must be booked in advance. Partial loads are either picked up without booking according to contract at regular pick-up times or picked up by special order from customers who have no contract.

For goods that are booked, the time interval between booking and pickup is extremely short: only 0.9 days on average. There is no significant difference between manufacturers and wholesalers with regard to this time interval. The total time interval between a received order and shipment, the order cycle, is 10.7 days which means, in principle, that the sender could book transport 9.8 days earlier. The average order cycle of wholesalers is 3.2 days, compared to 17.9 days for manufacturers which, similarly, gives 2.3 and 16.0, respectively, as potential increases in advance booking. These results indicate that it is possible for the sender to book transport earlier.

This view is also supported by responses to the question: "When in the order cycle does booking ordinarily occur?". Table 2 shows how the companies' booking is distributed among various elements in the order cycle (% of companies). The proportion of companies that book late in the order cycle is as high as 64% and significantly higher for wholesalers (71%) than for manufacturers (58%), while only 17% may be said to book early. Wholesalers have a somewhat more U-shaped distribution of booking over the order cycle which is significantly different from that of the manufacturers.

Table 2. Location of booking in the order cycle (% of companies)

Location of booking in the order cycle	Category of sender		
	Wholesaler	Manufacturer	All
When order is received	15	8	11
When time of delivery is determined	6	6	6
When order is confirmed	3	2	2
When picking list is compiled	3	9	7
When packing order is compiled	3	14	9
When goods have been packed	68	52	60
When goods are on loading deck	3	6	4

Earlier booking

For full loads, senders indicate that 34% of the present quantity of booked goods could be booked one day earlier. The corresponding figures for 2 and 3 days are 13% and 3%, respectively. Small wholesalers and manufacturing companies account for the highest share of such temporal flexibility. These companies indicate that 40% and 39%, respectively, of goods now being booked could be booked one day earlier. The possibilities for booking earlier can also be explained in terms of the type of activity conducted by both sender and recipient, in terms of trade or manufacturing, as shown in Table 3. The table shows the average proportion of goods weight that could be booked one day earlier (%) for various combinations of sender and recipient. The four cell averages are significantly different, and the average proportion for wholesalers as senders is significantly higher than that of manufacturers as senders.

Table 3. Average proportion of goods weight (%) that could be booked one day earlier for two categories of recipients

Sender	Recipient	
	Manufacturer	Trading Company
Manufacturer	22	32
Wholesaler	35	42

50% of companies shipping out goods require no incentives at all from carriers for booking earlier, while 37% indicate that they would book earlier if given a lower price. It is mainly the larger companies that require no incentives, while small companies to a greater extent require lower prices than do the upper strata. Thus, carriers could achieve a large share of the temporal flexibility resulting from earlier booking with no significant sacrifice.

The companies that could not book earlier indicated that this was primarily due to their not knowing when the order would be ready for shipment. This uncertainty is often caused by internal problems in planning the flow of orders. Other important reasons are that customers order late and that there is uncertainty concerning the size and weight of the order. Several companies indicated that earlier booking presupposed, or would be facilitated by, an on-line system of communication with the carrier.

Regarding part loads, i.e. goods that are not booked at present, senders indicate that 51% of this quantity of goods could be booked one day before pickup and 11% two days before pickup. As an incentive, 53% would require a lower price, while 39% would ask for no incentive at all. A significantly higher proportion of wholesalers than manufacturers call for lower prices in compensation.

Forecast information

At present, 19% of the senders submit forecast information on their shipping needs to their carriers. The remaining companies believe that it would be possible to submit forecasts on their transport needs and 66% would do so without compensation, while the others would require lower prices or higher transport quality. As an average for all companies, including those that already submit forecasts, advance forecasting could be increased by 2.1 days. Among wholesalers the proportion of medium sized and large companies that indicate that they could submit early, non-binding information on their planned shipments without compensation is significantly higher than the corresponding proportion of small companies. Among manufacturing firms, there is no corresponding pattern. Several companies have indicated, however, that this type of information transfer would require on-line data communications with the carrier and, in certain cases, the development of internal computer-based information systems.

The chances of increasing carriers' lead time for planning through earlier booking and early, non-binding information on planned shipments, such as preliminary booking, must be considered good. Carriers could obtain earlier information on a large percentage of goods without having to offer any real incentives.

Flexible pickup

Responses to the questionnaire indicated that 18% of the senders could accept having the carrier select the pickup time within a 2-day interval for each shipment, which corresponds to 10% of the total weight of goods on the market. If this time interval is extended to three or five days, the number of senders drops to 6% and 3%, respectively, corresponding to 4% and 1% of the total weight of goods on the market.

Companies with more than 50 employees are least willing to accept flexible pickup and this is true about both wholesalers and manufacturers. One probable reason is that more of the larger companies work with time-managed shipments. Table 4 provides additional perspective. The table shows that the chances of a sender offering flexible pickup vary for different combinations of sender and recipient.

Table 4. Proportions of shipments (%) that could offer flexible pickup within a two-day interval for various combinations of sender and recipient

Sender	Recipient	
	Manufacturer	Trading Company
Manufacturer	14	18
Wholesaler	16	20

57% of the companies that could accept some form of flexible pickup indicated that they would require a lower price in compensation. 42% would require higher transport quality, while 27% would require nothing at all. It was indicated that both these categories assume increased cooperation with carriers. In addition, it is mainly the small companies that called for lower freight prices as a requirement for accepting a flexible pickup arrangement.

The reasons most frequently given by companies that cannot accept flexible pickup at all are small stocks, a desire for short lead times, a need to have the customer receive goods the day after the order is placed, customer-controlled JIT deliveries that make flexibility impossible, use of a transport concept based on fixed departures and fixed arrivals, presence of personnel in the warehouse only during pickup and delivery, existing flexibility in the overall order cycle has been reserved for future purposes, etc.

Some companies justify their negative attitude by saying that flexible pickup could make shipping more expensive, due to coordination problems between sender and carrier. However, this need not be the case if the transport is performed on exchangeable load carriers and if the carrier can inform the customer on short notice using suitable information technology, i.e. flexibility-adapted technology.

Flexible deliveries

The study shows that there is considerable scope for the use of flexible times for the delivery of goods to the recipient. Sending companies estimate that 24% of those receiving goods would accept having the carrier choose the time of delivery within a 2-day interval determined by the sender. This corresponds to 10% of the total weight of goods on the market. If the interval were extended to include 3 days, the number drops to 5%, corresponding to 4% of the total weight of goods.

Among wholesalers, it is primarily small and medium-sized companies that believe that their customers could accept this. The pattern is not as clear among manufacturing firms.

The reasons most frequently given by senders as to why their customers cannot accept flexible deliveries are that they want to be able to plan their own activities, that they order late, that their needs are acute (spare parts), or that they are in a sensitive competitive situation. Some negative respondents add that they could work with flexible deliveries if they had access to an information system that identified which shipments could use flexible delivery. For them the problem is more shipment-specific than customer-specific. Some senders justify their negative response by stating that flexible deliveries require that the recipient be able to mobilize his receiving capacity at unpredictable times. However, this need not be the case if technology adapted for flexibility is used, as indicated in the previous section.

Interactive booking

Respondents were also asked about their view on an alternative system of cooperation between sender and carrier in which interactive booking is used. In this system the sender gives the carrier the latest possible time a shipment must reach the recipient, a proposed pickup day, and an interval around the proposed day, from the earliest day to the latest day. It is assumed here that in most cases the carrier will accept the proposed pickup day, but in certain cases he will propose another day within the pickup interval. The interactive booking system is illustrated in Figure 3.

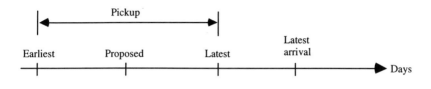

Figure 3. Interactive booking system for shipments

35% of the companies believe it would be possible to work together with carriers in the domestic, line-based trucking business using a system of the kind described above. These companies estimate that 50% of their goods could be booked interactively, which corresponds to about 8% of the entire weight of goods on the market. They believe the maximum average pickup interval could be 2.5 days. Many of them are medium-sized wholesalers.

Negative comments from respondents on interactive booking are that expectations of domestic trucking lines were too high to permit use of this system, that the system would mean that the sender must have storage space for goods that had been booked but had not yet been picked up, that the system would entail additional administration of shipments, and that the sender must coordinate delivery with the customer.

Possible barriers

There are probably various reasons why the potential temporal flexibility has not yet been utilized. The survey and the preliminary study revealed several possible contributing factors:

- Genuine ignorance, meaning that actor A does not know that the flexibility which he can create is of value to actor B or that actor B does not know that actor A can offer such flexibility.

- There are no information systems for the operative coordination of A's and B's activities under a flexible cooperation system.

- It is difficult under current contractual conditions to distribute between A and B the costs and benefits that a flexible system would entail.

- B is a facilitating actor in many marketing channels and must consolidate the flexibility of several different customers.

If these factors have presented some kind of barrier to the utilization of interorganizational flexibility in marketing channels in the past, they are rapidly decreasing in significance. Companies' IT capacity and IT capabilities are increasing, and flexible manufacturing systems will also create possibilities. Acceptance of interorganizational solutions such as partnerships and strategic alliances is on the rise and there is more and more interest in industry-wide efforts toward greater channel efficiency by means of standardization and cooperative measures. Examples of the latter include the rapidly growing interorganizational cooperation in North America and Europe on Quick Response (QR) and Efficient Consumer Response (ECR) (see, for example, Stern et al., 1996).

CONCLUSIONS AND DISCUSSION

Conclusions of the empirical study

This study shows that the customer order cycle among members of logistics channels has a structure that will permit increased temporal flexibility in transport activities between these members and the next level without any degradation in service. This executive flexibility can be utilized by carriers in the channels to improve the efficiency of the transportation function. The study was aimed at shipments from manufacturers and wholesalers to the next level in these channels. The results show that pickup at the sender and delivery to the recipient can occur flexibly within a time interval instead of at a definite point of time for a considerable share of the flow of goods measured in shipments or weight units. Carriers can utilize this flexibility to pick up and deliver goods in accordance with a time pattern that will minimize variations in the flow of goods from day to day and, in particular, the maximum flow of goods, which is the main determinant of transport capacity. There is also a good potential for shippers to increase their advanced booking of shipments and to submit early forecasts of their transport needs. This planning flexibility will give carriers longer lead times, which will allow them to utilize their resources more efficiently. If carriers can utilize the various types of flexibility that are available or that can be created, they will have a potential for increased resource utilization with regard to vehicles, personnel, handling equipment, and terminal space.

Methodological comments

The available flexibility is probably greater than indicated by the empirical study. This is because respondents, those who are responsible for transport at manufacturers and wholesalers, in interviews, will mentally restrict their thinking to the possibilities for flexibility within the framework of the present distribution of responsibilities and planning routines within their own organizations and with their carriers and within the framework of the technology they are using at present. A deeper penetration of the potential for flexibility, given a more flexibility oriented planning system or alternative, flexibility oriented technologies, cannot be expected in an interview.

The shipper's information system

For a given shipper, the degree of flexibility will be a shipment-specific concept and it will depend on the products which the shipment contains, the customer's priorities, and the shipper's capacity. The estimated flexibility per shipment can be based on information sources already present at the company and it can be administratively coupled to existing activities in the order cycle which, in chronological order, may be order booking, credit check, determination of delivery date, manufacture or picking, packing, and placement on the loading dock. If the sender lacks information on the customer's ability to accept flexible delivery, then this information could be routinely included in order information from the customer. Effective planning requires that the sender's internal information system be computer based unless the number of shipments is low, which should present no great difficulties. This was indicated by the respondents in the empirical study.

Communication between shipper and carrier

Shippers have a good opportunity to offer carriers increased advance planning and flexible pickup and delivery with no degradation in customer service. If the potential is to be utilized by carriers, then, in many cases, a system of computer-to-computer communication between shipper and carrier will probably have to be introduced. This is particularly true of companies with many shipments, as well as for the transfer of forecast information and for interactive booking. Introducing a computer communications system between shippers and carriers would also bring about a number of other advantages that would help to reduce costs or increase service. Shippers cannot be expected to accept investments in a communications system that would tie them to a certain carrier. Consequently, utilization of potential flexibility would be facilitated if an industry-wide EDI standard were developed.

Partnership between shipper and carrier

Full use of potential flexibility to increase resource utilization in the transportation system presupposes that the present contract-based relationship between shipper and carrier be further developed in the direction of partnership. Medium-sized and large senders can offer carriers a certain amount of flexibility with no special consideration, but full utilization of this potential will require investment and operating costs that would have to be compensated either in the form of lower prices or higher transport quality from carriers. Since carriers would also incur initial investment costs in communications and planning systems before realizing any benefits, a long-term relationship is needed between carrier and shipper so that costs and benefits would be divided between the two actors, i.e. a partnership would be created. Such a development is also in line with the trend toward increased interorganizational cooperation between members of logistics channels in combination with a different distribution of labor and responsibility, such as third-party logistics, which is a result of IT developments and increasing specialization.

Further research

This study has shown that interorganizational logistics flexibility is a promising field of research and that the system studied contains potential flexibility of such magnitude and nature that it would be of interest to continue following the research strategy proposed in the introductory section. Thus, it seems that further research into logistics flexibility could help to increase efficiency in marketing channels. The following areas seem ripe for research.

1. Simulation studies based on existing demand for transportation, showing what increases in resource utilization and cost effectiveness could be achieved by taking advantage of

potential flexibility under alternative ways of designing and managing the transportation system.

2. Aspects of design and behaviour of intra- and interorganizational information systems for extracting and communicating information on flexibility.

3. Development of support systems for decision-making for optimum utilization of information on flexibility in resource management at transportation companies.

The empirical study deals with temporal flexibility in an ordinary class of transport-related activities in logistics channels. The set of concepts is general, however, and the approach may be applied to other flexibility dimensions and other logistics activities, as well as to distribution activities in transaction channels. Since the phenomena that make it possible to create and utilize flexibility in practice within marketing channels are increasing in number and importance - such as time control and standardization of activities in marketing flows to recipients with differentiated needs - there is a need for more research on flexibility, both as a complement and an alternative to the rather extensive research that is focusing on rigidity in marketing channels.

ACKNOWLEDGEMENT

The author wishes to thank the Swedish Communication Research Board for funding this project and Dr. Peter Rosén, Dep. of Business Adm., Gothenburg University for cooperation in the data collection phase of the study.

REFERENCES

Aaker, D.A. and Mascarenhas, B. (1984). "The Need for Strategic Flexibility", Journal of Business Strategy, Vol. 5, No. 2, pp. 74-82.

Adler, P.S. (1988). "Managing Flexible Automation", California Management Review, 30,3, pp. 34-56.

Andersson, P. (1991). "Analyzing Distribution Channel Dynamics: Loose and Tight Coupling in Distribution Networks", European Journal of Marketing, Vol. 26, No. 2, pp. 47-68.

Ansoff, H.I. (1975). "Managing Strategic Surprise by Response to Weak Signals", California Management Review, 8,2, pp. 21-33.

Billesbach, T. J. and Hayen, R. (1994). "Long-Term Impact of Just-in-Time on Inventory Performance Measures", Production and Inventory Management Journal, Vol. 35, No. 1, pp. 62-67.

Buzacott, J.A. (1982). "The Fundamental Principles of Flexibility in Manufacturing Systems", Proceedings of the First International Conference on Flexible Manufacturing Systems. Brighton, UK.

Bowersox, D.J., Cooper, M.B., Lambert, D.M. and Taylor, D.A. (1980). Management in Marketing Channels, New York: Mc Graw-Hill.

Carter, P. L., Melnyk, S.A. and Handfield, R.B. (1995). "Identifying the Basic Process Strategies for Time-Based Competition", Production and Inventory Management Journal, First Quarter, pp. 65-70.

Das, T.K. and Elango, B. (1995). "Managing Strategic Flexibility: Key to Effective Performance", Journal of General Management, Vol. 20, No. 3, Spring, pp. 60-75.

Daugherty, P.J. and Pittman, P.H. (1995). "Utilization of Time-Based Strategies", International Journal of Operations and Production Management, Vol. 15, No. 2, pp. 54-60.

de Meyer, A., Nakane, J., Miller, J.G. and Ferdows, K. (1989). "Flexibility: The Next Competitive Battle: The Manufacturing Futures' Survey", Strategic Management Journal, 10,2.

Eppink, D.J. (1978). "Planning for Strategic Flexibility", Long Range Planning, Vol. 11, pp. 9-15.

Evans, J.S. (1991). "Strategic Flexibility for High Technology Manoeuvres: A Conceptual Framework", Journal of Management Studies, 28:, January, pp. 69-89.

Frazier, G., Spekman, R. and O'Neal, C. (1988). "Just-in-Time Exchange Relationships in Industrial Markets", Journal of Marketing, (October).

Gadde, L.-E. and Håkansson, H. (1991). "Analyzing Change and Stability in Distribution Channels - A Network Approach", in Axelsson, B. and Easton, G. (Eds.), Industrial Networks - A View of Reality, London: Routledge.

Gattorna, J. (1978). "Channels of Distribution Conceptualizations: A State-of-the-Art Review, European Journal of Marketing, Vol. 12, No. 8, pp. 471-512.

Gerwin, D. (1993). "Manufacturing Flexibility: A Strategic Perspective", Management Science, Vol. 39, No. 4, April, pp. 395-410.

Gupta, Y.P. and Goyal, S. (1989), "Flexibility of Manufacturing Systems: Concepts and Measurements", European Journal of Operational Research, 43, pp. 119-135.

Hall, R. and Tonkin, L. (Eds), (1990). Manufacturing 21 Report: The Future of Japanese Manufacturing, Association for Manufacturing Excellence, Wheeling, IL,

Harrigan, K.R. (1980). "The Effect of Exit Barriers on Strategic Flexibility", Strategic Management Journal, 1,2, pp. 165-176.

Harrigan, K.R. (1985). Strategic Flexibility: A Management Guide for Changing Times. Lexington, Mass.: Heath.

Hise, R.T. (1995). "The Implications of Time-Based Competition on International Logistics Strategies", Business Horizons, September-October, pp. 39-45.

Hutchinson, G.K. (1973). "Flexible Manufacturing Systems", Industrial Engineering, 5, p.10.

Håkansson, H. and Johansson, J. (1992). "A Model of Industrial Networks", in Axelsson, B. and Easton, G. (Eds.), Industrial Networks - A View of Reality, London: Routledge.

Jensen, A. (1990). Combined Transport. Systems, Economics and Strategies, Stockholm: Swedish Transport Research Board (TFB) and Allmänna Förlaget.

Krijnen, H.G. (1979). "The Flexible Firm". Long Range Planning, April.

Mallen, B. (1973). "Functional Spin-off: A Key to Anticipating Change in Distribution Structure", Journal of Marketing, Vol. 37, July. pp. 18-25.

Mandelbaum, M. and Buzacott, J. (1990). "Flexibility and Decision Making", European Journal of Operational Research, January, 44,1, pp. 17-27.

Merkhofer, M.W. (1977). "The Value of Information Given Decision Flexibility", Management Science, Vol. 23, No. 7, March, pp. 716-727.

Pye, R. (1978). "A Formal Decision - Theoretic Approach to Flexibility and Robustness", Journal of the Operational Research Society, Vol. 29, No. 3, pp. 215-227.

Stalk, G. and Hout, T.M. (1990). Competing against Time, New York: The Free Press.

Stern, L.W., El-Ansary, A.I. and Coughlan, A.T. (1996). Marketing Channels, Upper Saddle River, New Jersey: Prentice-Hall.

Weick, K.E. (1976). "Educational Organizations as Loosely Coupled Systems", Administrative Science Quarterly, Vol. 21, pp. 1-1

5

QUANTIFYING TRANSPORT QUALITY IN LINE-BASED ROAD TRAFFIC

Johan Hellgren
Kenth Lumsden

ABSTRACT

This paper provides a model for quantitative measuring of transport quality in line-based road traffic. The model primarily focuses on the operational level, i.e., the production of an individual transport. The model takes different aspects of transport quality into account; different quality components are important to different actors (e.g., shipper, carrier or forwarder), and different components are relevant in different situations (e.g., core (production-related) and shell (consumption-related) components).

A survey has been sent to a sample of transport buyers; Swedish manufacturing companies with more than 50 employees. The questionnaire used included sections on the relative importance of different performance factors and open questions on critical aspects of a transport, but also questions on types of operations and the structure of the transports of the respondents. The transport producers' view has been explored by a case study in a transport company, in which the transport company's opinions about the aspects that were included in the survey were established.

The most important quality aspects from the surveyed sample's point of view are, irrespective of their type of operations, transport time and regularity. This finding paired with corresponding case study results suggests that continuous monitoring of the most important of the quality components is possible by using automated data capture methods. Another finding of the survey was that 'soft' factors such as professionalism and courtesy are considered to be very important. While these aspects lend themselves less easily to continuous monitoring this finding does, however, emphasise the importance of a customer-oriented culture both in the transport company itself, but also among its sub-contracted hauliers.

Keywords: Transport quality, customers' opinion, performance measurement, line-based road traffic, information technology

INTRODUCTION

The quality concept and quality improvement have long been focal points of attention in the manufacturing industry. It is only in later years that this trend has spread to the transport industry. Due to this, no universally accepted definition of transport quality exists and, because of the lack of definitions, no general model of transport quality has been developed.

What does exist, are different descriptions of transport quality, often derived from definitions of service quality in general (e.g. Hopkins et al. 1993). However, due to their extensive generality, these definitions rarely permit measuring the quality level of a transport service without major adaptations and developments. What also exist, are different measures and measuring methods developed and used by different transport companies. While being quantitative, these measures are rarely based on existing theory and do not necessarily focus on the quality aspects that the customers consider to be important. There is hence a need for a quantitative quality model that is theoretically founded and contains components that can be continuously measured, i.e., measured on every single transport. In order to ensure that the model contributes to increased customer satisfaction, the customers' priorities should be considered when inserting quality measures into the model. It also must be stressed that any quantitative measures are highly dependent on the availability of information technology to ensure satisfactory data capture. This is especially true in the transport industry, since its wide-spread production facilities (load carriers and terminals) do not lend themselves to centralised data capture.

TRANSPORT QUALITY

Scope of the quality concept

Production and consumption of a product or service affect a considerably larger system than merely the producing and consuming organisations. There are hence quality aspects to be applied to a larger scope than the fundamental, i.e., the core quality of the product or service. An example of this is to consider the environmental aspects of the production of a product or service. The widening scope of quality aspects is shown in Figure 1.

Figure 1. The widening scope of quality aspects (Wadsworth et al. 1986)

On the company level, it is necessary to demarcate those aspects that are to be included in the quality concept. This demarcation varies depending on the business type. A single company has normally less possibility to exert influence over the quality aspects on the periphery (see Figure 1). One example of this is the environmental aspect, which must be controlled on a national level at least. However, the global aspects can be made more fundamental by, for instance, levying taxes on excessive emissions or waste. The quality aspects of immediate interest to a single organisation must be confined to those that can be controlled by the organisation itself. These aspects are usually confined to the four inner circles (see Figure 1). These circles cover the design, production and delivery of the product, and the production system as a whole.

Stage in the Product's Life Cycle

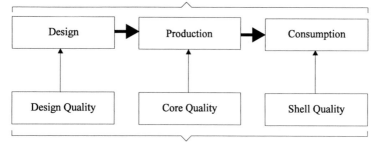

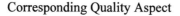

Corresponding Quality Aspect

Figure 2. Scope of the quality concept within a company (Hellgren, 1994)

As has been indicated, the quality of the design of the product is called design quality. The design quality level is a measure of the fundamental ability of the product or service to fulfill the customer's requirements. It also covers areas such as how easily the product can be produced according to the specification of the design (Wadsworth et al., 1986). This means that design quality often is more important when producing products than services, since the design of a product generally is more complex than that of a transport or other service.

The production of the product or service is covered by the term core quality. Core quality in its fundamental form can be said to be the product's or service's conformance to specification (Wadsworth et al., 1986). The specification is the result of the design process and covers things like, e.g., measurement tolerances or transport times. Note that core quality does not relate directly to customer satisfaction, since a product can conform exactly to specification and still be considered as unsatisfactory by the customer, provided that the specification fails to meet the customer's requirements.

The delivery or consumption phase is covered by the shell quality concept. Shell quality covers the customer's opinion on the quality level of the company for all those aspects that are not directly related to the product or service, e.g., delivery time or response time. For a service, which by definition is produced and consumed simultaneously, this means that both core and shell quality can be linked to the production system. For a product, the shell quality is more often the responsibility of the sales and marketing, and after-sales departments.

Service quality and quality gaps

The most fundamental definition of service quality is the extent of discrepancy between customers' expectations or desires of service and their perceptions of the service they actually receive (Zeithaml et al. 1990). This definition was broken down into a model of service quality in the form of quality gaps. A quality gap can be defined as the difference between what the service-providing company and its customer expect, both concerning the nature of what is expected and the performance for those components. One example of the first is the difference between the customer wanting consistent transport times while the transport company thinks that fast transport times are required. An example of the second is the difference between the customer requiring 98% of the shipments to be delivered on the specified times while the transport company delivers only 95% of the shipments on time. This means that the quality gap can be measured either on a nominal or on an interval scale. Hopkins et al. (1993) have adapted the service quality gap model to fit the transport industry. This model is shown in Figure 3.

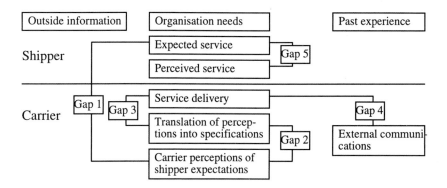

Figure 3. Service quality model adapted to transportation (Hopkins et al. 1993)

The first gap represents discrepancies in expectations between shippers and carriers, i.e., between what the shippers think and the carrier thinks that the shippers think. This is a gap of expectations. Clear communication between shipper and carrier reduces this gap. The second gap represents misinterpretation, by the carrier, of shipper expectations into quality specifications. This is a gap of interpretation. The third gap represents service quality performance inconsistent with the specifications, i.e., failures in the production system of the carrier. This is a gap of performance. Gap four is the discrepancy between what is actually delivered and what is communicated to the shipper as delivered. This is a gap of communication. Gap five is the aggregate sum of the four previous gaps. Different gaps possibly can cancel each other out but, as long as they are treated separately, it may be assumed that gaps 1-4 may safely be aggregated, although the unit of measurement of course in many aspects will be subjective.

Major issues that arise when using this model are what data it should be based on and its impact on customer satisfaction. Bolton and Drew (1991) state that the aggregate expectation gap is a major driver of customer satisfaction. This suggests that it is useful to implement the model. However, how to measure customer expectations still remains to be resolved. Clow and Vorhies (1993) argue that expectations are stable in the short term. Expectations need therefore not be measured continuously but may instead be assessed at regular intervals. They also elaborate on whether expectations should be measured before or after the consumption experience, which they relate to the possibilities of obtaining a repeat purchase. In the transport industry, where customers rarely are individual consumers, this aspect is, however, of less importance, since some degree of repeat purchases will be obtained in any event.

Measuring customer expectations and using the service gap model will result in a set of gap sizes. There are no fundamental obstacles to using this model in a transport company. What seems to differ between transports and services in general is more what components constitute the quality of a transport or service, rather than any major differences in approach. If anything, transports have a larger core (physical) content than pure services and would therefore be more suited to continuous monitoring of quality as is common in the manufacturing industry. It is important to note, however, that a large portion of the reasons for poor perceived quality is due to inadequacies in communication, rather than inconsistent performance, which is important to consider when contemplating the impact of performance-enhancing measures.

Quality in transport services

A transport can be described as a flow of materials between two organisations. A general definition of logistics can be used as a basis for defining the quality of a flow of materials (Strand & Lumsden, 1989):

"The right product, in the right quantity, in the right condition, at the right place, on the right time, to the right customer, at the right cost." (Shapiro & Heskett, 1985)

All aspects of this definition can be applied to a single transport. However, transport quality can also be described using other dimensions. Among these are (Jensen, 1990):

- Frequency, i.e., the number of departures per time unit;

- Transport time;

- Regularity, i.e., the ability to conform to an agreed time schedule;

- Goods comfort, i.e., protection against shocks, climate and chemicals;

- Transport security, i.e., protection against theft;

- Controllability, i.e., possibilities to track consignments underway;

- Flexibility, i.e., adaptability to changes in the flow's volume, direction and content;

- Detachability, i.e., if the transport can be performed using less of the consignor's or consignee's equipment;

- Expansibility, i.e. the ability to expand by discrete steps.

Of these dimensions, all except the frequency, flexibility and detachability are relevant to a single transport. These dimensions focus on the performance and capabilities of the transport system as a whole. It is also notable that Jensen's dimensions are primarily connected to the core quality concept. Detachability also may be considered as relevant to a single transport. Due to the quantitative focus of this paper and the difficulty to quantify these components, these components nevertheless have been disregarded.

Dividing transport quality into dimensions as above is one way of making this concept more concrete than is achieved by merely using the Heskett & Shapiro general logistics definition. To further the concretisation of the transport quality concept, it is therefore useful to establish the relations between the transport quality dimensions and the components of the logistics definition.

Logistics component   **Transport quality dimension**

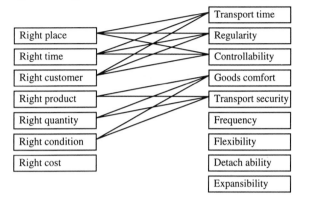

Figure 4. Relations between logistics components and transport quality dimensions (adapted from Heskett & Shapiro,1985 and Jensen, 1990)

The right product in the right quantity and condition can be linked to the goods comfort and transport security dimensions, since only damaged or stolen goods are permanently lost; misdirected transports can be corrected. At the right place on the right time can be linked to the transport time, regularity and controllability dimensions. The right time is achieved through fast and consistent (regular) transport times. Failure to reach the right place can be avoided by, for instance, using a good tracking system.

Christopher et al. (1979) define service as 'the conditions of response to orders'. These conditions consist of the terms under which the product flow is provided by the distribution system. These terms include the trade-off between the inputs, i.e., costs, and outputs, i.e., the functional objectives, of the distribution system. Transport quality can be seen as part of the functional objectives of the distribution system. Given this, it can be stated that the cost element of the general logistics definition fits into the transport quality concept. However, "at the right cost" is not covered by any of Jensen's transport quality dimensions. The right cost from the customer's point of view is the agreed price of the transport, since that is what he or she considers the transport service to be worth. However, from the transport company's point of view the right cost is not necessarily equal to the agreed price. Instead, the right cost is the lowest cost that still enables the company to produce a transport service that fulfills the other quality requirements. This argument is valid even if the transport service is produced by several independent organisations. In this case, i.e., when a company that produces a transport service sub-contracts part of it to other firms, the right cost from the company's point of view for each portion of the transport is equal to the price agreed with the respective sub-contracted firms, whereas each of the sub-contracted firms strives to minimise its costs while still fulfilling the other quality requirements. This view corresponds to the analogous case of an internal production process, for which Ishikawa (1985) states that 'The next process is your customer.'

THE CUSTOMERS' VIEW ON TRANSPORT QUALITY

In order to determine the true focus of the transport customers' quality focus, a questionnaire was sent to a sample from the PAR (Swedish Postal Service) database representing 50% of Swedish manufacturing companies with more than 50 employees, and a person listed as responsible for logistics. The questionnaire was addressed to this person in each firm. In total, 210 companies received the questionnaire and 159 responses were received, equivalent to a response rate of 75.7%. The questionnaire included one part where the respondents were asked to rate the quality dimensions of the previous section on a four-grade scale ranging from decisive importance to unimportant. In addition, the respondents were given the opportunity to give their opinions of significant quality aspects for different parts of a hypothetical transport. The objective of this part of the questionnaire was to ascertain whether structuring transport quality along transport industry-defined lines was consistent with the customers' view on transport quality.

Another part of the questionnaire dealt with the quality measuring activities of the respondents. The purpose of this was to establish how common quality measuring is today and how important monitoring transport quality is considered to be.

The mail survey resulted in quality dimension importance ratings as depicted in Table 1 below. The responses have been divided by industry type. Only responses representing a minimum of 50% of the total responses of the respective industry types that attach decisive or major importance to the respective dimensions have been included.

As can be seen, the components considered most important by the respondents are regularity and transport times. This means that what the customers above all others want is a fast and reliable transport.

The results of the free format questions can best be described by depicting the customers' requirements on each part of the transport. In the questionnaire the transport was divided into booking, pick-up, transport, delivery and administrative contacts.

When booking a transport the respondents' foremost requirement is for a high degree of automation, i.e., contacts only by way of EDI. This is especially true for larger companies. For non-routine bookings that are impossible to make through EDI, the respondents generally want a special person to handle all business from one customer. This is hoped to facilitate communication and yield better advice since that person is expected by the respondents to have knowledge of the needs of the individual customer. Lead times no longer than one day are required by most respondents.

Table 1. Rating of quality dimension importance by industry type

Industry type	Degree of importance										
	Decisive			Major							
	A	B	C	A	C	D	E	F	G	H	I
Chemical	67	67			67	67	67	67			50
Construction	67	67	50				75	63		63	
Electronics		50		58	52	60	55	52			
Manufacturing		70		66	100	66		100			
Process	54	64					79	85	71	67	
Subcontractor		50		54		56					

Quality component key:
A = Transport time [%] F = Goods comfort [%]
B = Regularity [%] G = Transport security [%]
C = Flexibility [%] H = Detachability [%]
D = Frequency [%] I = Expansibility [%]
E = Controllability [%]

The chief requirement on the pick-up is, predictably, that it is made on the specified time. Other than that, the consensus is that the personnel and equipment of the transport company must be tidy and professional-looking and that the driver must participate in loading the vehicle to ensure that the goods are secured safely in place.

During the transport phase, the predominant requirement is that the customer should be promptly informed of deviations from the agreed schedule. Most respondents were also adverse to transhipment. However, since transhipment is a necessary activity in any line-based network, this requirement may be interpreted as a reluctance to accept goods damages.

When delivering the goods the most common requirement is that the driver must act professionally, as he is perceived as the representative of the sender. More concretely, many respondents stated that the driver must ensure that the goods are delivered to the right person, and not to anyone that happens to be nearby as the driver arrives. Several respondents also wanted the driver, together with receiving personnel, to make a preliminary inspection of the goods to immediately identify goods damages. This is an inherent problem for most of the service industry, since in a manufacturing firm the people that meet the customers are the best educated of the company (e.g., salespeople), but in the service industry they are often the least educated (e.g., drivers) (Persson & Virum, 1995).

The remarks on administrative contacts between transport company and customer focused mainly on invoicing. Invoices comprising several, clearly identifiable, transports were desired by most respondents. Routine invoices should be transferred by way of EDI. One general requirement was that all documents must be easily connectable to specific transports. Any deviations from previous agreements should be clearly recognisable. As was also true for booking, most respondents want the same person to handle all administrative contacts with a customer.

The results of the free-format questions are summarised in Figure 5. The figure depicts important, quality-related aspects of different parts of a transport and is based on all responses, irrespective of industry type.

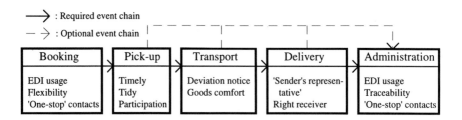

Figure 5. Summary of customer-emphasised transport quality elements of a transport

The resulting scope of transport quality on a company level will thus comprise the physical transport, customer relations and handling and the flexibility of the transport system. This focus is roughly equivalent to the core, shell and, to a lesser extent, design quality aspects.

CONCEPTUAL MODEL OF TRANSPORT QUALITY

So far, the relevant transport quality aspects and the general principle, that the quality improvement procedure should always begin by finding out what parameters the customers emphasise and conclude by measuring the actual level of customer satisfaction of a specific action, have been established. At this point, it is necessary to give the concepts of design, core and shell quality a more concrete content in order to obtain a conceptual model that can serve as a definition of transport quality.

It was previously stated that the quality scope in a transport company comprises core and shell quality and possibly also design quality. These general terms need to be adapted to fit in a transport company. They also need to be broken down into different components to be meaningful. When constructing these components, it is essential to include the customers' view and not only focus on factors of purely internal importance. The resulting quality components can therefore be regarded as belonging to different categories depending on whether they are primarily relevant to the customer's impression of the transport or instead relate to the efficiency of the transport company.

Design quality components

Design quality is generally of limited interest in a transport company. This is mainly due to the design of a transport service being fairly straight-forward. Therefore, it is rarely possible to better satisfy the customers' demands by redesigning the transport. The possible exception to this is to decrease the transport time by employing a faster means of transport. This action can, however, be seen as designing an entirely new transport service. Design alterations on a lower level, for instance re-routing, are normally done as corrective actions in response to changes in the environment during the production phase. Generally, it is hence possible to demarcate design quality in a model for measuring quality in a transport company, which is also supported by the mail survey responses as previously described.

Core quality components

For a transport company, core quality means all the quality parameters that can be connected to the physical transfer of the goods (Lumsden, 1989). This means that core quality is the category that is easiest to measure for a single transport, since some form of physical transport of goods is part of every transport service. Of Jensen's transport quality dimensions, transport time, regularity, goods comfort, transport security and controllability can be categorised as core quality dimensions. Of these, customer emphasis lies, according to empirical studies, primarily on transport time, goods comfort and transport security (Wilson, 1980; McGinnis, 1990). The mail survey responses mention accuracy regarding transport time as important to the customers. This factor corresponds to the regularity dimension, since a transport that is performed according to specification is, by default, accurate regarding transport time.

Measuring these dimensions would cover those aspects of transport quality that the customers prioritise. However, further dimensions must be measured. For instance, it is possible to maintain high performance levels for the customer-focused dimensions while operating inefficiently within the transport company. Internal dimensions must therefore be measured as well. Critical indicators of internal efficiency are primarily cost accuracy and number of unscheduled activities, since a company that has its costs under control and performs its activities according to plan can be considered as efficient.

Shell quality components

The term shell quality covers how the customer perceives the transport company's attitude towards its task and clients (Lumsden, 1989). Monitoring of shell quality is often undertaken by the customers of transport companies. One example of this is the computer company Nixdorf, which measures its transporters for, among others, invoicing, complaints, response time, financial strength and innovativeness (Bowan et al. 1990). Another example is the chemicals company Air Products and Chemicals, which in a weighted form measures its railway transporters for, among others, toxic emissions, end-consumer complaints and own complaints (Gordon, 1989). This is partly supported by the mail survey responses, according to which attention should primarily be placed on the responsiveness of the transport company. Other than mere response times for telephonic or other direct contacts it is, however, difficult to measure shell quality components in the day-to-day operations. Instead, these aspects are better suited for measuring at periodical audits.

Conceptual model

In the previous sections, the transport quality concept was structured. To construct a conceptual model of transport quality on a company level it is also, however, necessary to determine when in the transport production process the different quality components are relevant. From a production process point of view, the completion of the transport service can be divided into a production and a consumption phase. Since a transport by definition is consumed simultaneously with its production, this division needs to be qualified. This can be done by defining the production phase as comprising the transport legs before the final leg to the goods recipient, i.e., all sub-processes not directly noticeable by the transport customer. The consumption phase would comprise those processes that are noticeable by the transport customer, i.e., transport legs to and from recipient and sender, and all administrative contacts with the customer. This means that the production phase will roughly be covered by the core quality concept, while the consumption phase will be covered by the shell quality concept. By linking the quality components to the transport production process it is then possible to obtain a conceptual model of quality in a transport company.

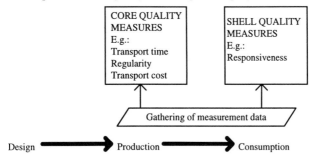

Figure 6. Conceptual model of transport quality on the company level

This model can be seen as a basic framework for monitoring and improving quality in a transport company. To be usable in practice for quality monitoring, the quality components must be quantified.

QUANTIFICATION OF THE CONCEPTUAL MODEL

Three principles have guided the quantification of the transport quality components. First, the measures should be usable in day-to-day operations, i.e., monitoring every transport should be possible. Secondly, the measures should be mergeable into a measure of the overall quality level of the transport. Thirdly, the measures should be able to serve as a basis for identifying where there is potential for improvement in the transport company. The first principle is the main reason behind the demarcation of the measuring model chiefly to core quality. The second principle is motivated by the need for an overall quality measure as a tool for marketing as well as for increasing the quality awareness within the transport company.

Based on our findings the measures have been grouped into three categories based on where they are primarily noticed. The first category, called external components, consists of performance parameters that are directly noticeable by the customers and hence of primary interest. The second category, called internal components, consists of measures that are unnoticeable by the customers, but give a clarification of the quality level internally in the transport company as a whole. The third category, called departmental components, consists of measures of the quality level of the different departments within the transport company. This category is motivated by the third quantification principle stated above.

External components

The timeliness of a transport can easily be defined and measured as the relative or absolute deviation from the delivery time agreed upon by the transport company and its customer. The delivery time agreement should preferably state if the delivery is to take place exactly on the specified time of arrival or if early deliveries are acceptable. The reason for this is that a consignment that is delivered early might not be required by the customer until the specified delivery time, thus requiring hard-to-spare storage space until that moment. Allowing for this, the relative accuracy of the transport can be quantitatively stated as

$$A_t = \left(1 - \frac{|T_r - T_c|}{T_c}\right) *100[\%] \text{, where}$$

A_t =   Accuracy with regard to time

T_r =   Real elapsed transport time

T_c =   Contracted transport time.

Consignment conformance is defined as the conformance between the contracted consignment content and the items that are in fact delivered. It can be measured in relative or absolute terms based on the value of the consignment or the number of items in the consignment. Any deviation from a perfect result for this parameter means that items have disappeared or been damaged underway. To determine the cause of non-conformance it is therefore necessary to divide it into three sub-categories:

- Items delivered but found to be damaged underway;

- Items missing due to misplacement underway, i.e., they are still in the transport system;

- Items missing due to theft or fraud, i.e., items that are permanently lost due to external factors.

In this way it is possible to determine if the physical strain underway, the planning system or alternatively the variation of transport times of the transport company or lack of transport security is the cause of non-conformance. An aggregated, relative quantitative measure of the consignment conformance would read

$$C_c = \left(1 - \frac{\left|I_d - I_c\right|}{I_c}\right) * 100[\%] \text{, where}$$

C_c	=	Consignment conformance
I_d	=	Number of items actually delivered
I_c	=	Total number of items in the consignment.

If the cause of non-conformance is to be determined, the absolute difference between Id and Ic needs to be divided according to the cause of non-conformance. If desired, Id can then be adjusted regarding cause category. In this way the impact of the individual causes of non-conformance on the consignment conformance can be calculated.

Internal components

So far, all performance aspects that are directly noticeable by the customer in the case of an individual transport have been covered. However, it is necessary to consider some factors that, though invisible to the customer, are of importance to the transport company.

The quality requirement "at the right cost" is by definition always met if one accepts the right cost as being equal to the agreed price of the transport. However, transport cost needs to be considered from an internal view as well, since the difference between the agreed price and the cost actually incurred is the operating profit of the transport company. It is therefore useful to carefully measure the transport cost for all transport services that are performed according to plan. The cost structure thus obtained will then determine the long-term minimum price required for the respective transport services to be profitable. It is important to confine these measuring activities to preplanned transports only, since the ideal transport system operates without unscheduled activities. Since establishment of such a system should be encouraged, one should not allow for such activities when determining the cost structure.

When measuring the transport cost it is important to attribute the different transport services with the right cost components, i.e., the cost of the resources that have in fact been consumed to produce a specific transport service. It will therefore be necessary to employ cost distribution methods such as the Activity Based Costing method to accurately determine the cost of the transport. Measuring the cost of all activities necessary to perform an individual transport is obviously difficult. However, this task is simplified if the transport company is certified according to the ISO 9000 standard, since this standard requires all activities to be extensively documented. This makes it possible to determine the cost of each activity, which in turn reduces the measuring activities to merely registering what activities have been performed in the production of the transport service. Given this ability to accurately determine the real incurred cost, a relative measure of transport cost accuracy can be given as

$$A_c = \left(1 - \frac{\left|C_r - C_p\right|}{C_p}\right) * 100[\%] \text{, where}$$

A_c	=	Transport cost accuracy

C_r = Real incurred transport cost

C_p = Projected transport cost.

One other occurrence that is invisible to the customer is a transport where unscheduled activities have been required in order for it to reach acceptable levels for its external quality components. Many such activities are an indication of deficiencies in either the planning system of the transport company or in the execution of the planned activities.

The unscheduled activities often need to be divided into homogeneous categories. These categories may comprise unplanned redirections of load carriers, employment of additional load carriers or employment of faster or more costly means of transport. However, a basic, relative, aggregated measure of the activity accuracy could be stated as

$$A_a = \left(1 - \frac{A_u}{A_s}\right) * 100[\%], \text{ where}$$

A_a = Activity accuracy

A_u = Total number of unscheduled activities

A_s = Total number of activities.

Note that this measure approaches zero as the number of unscheduled activities equals the number of scheduled activities. This implies that the activities in a transport are totally inaccurate when it is necessary to perform an unscheduled corrective activity for each scheduled activity.

The measure can easily be divided into separate components for each activity category. Analogous to the transport cost section above it is also possible to calculate the cost for each type of activity and hence include the activity accuracy in the transport cost accuracy measure.

Departmental components

The third category of measures comprises factors that describe the performance of individual departments of the transport company. These cannot be included in an aggregated measure of the overall quality level of the transport company, if the performance parameters they represent have already affected the aggregated performance of the transport company's departments, which is included in the internal quality components. These measures will naturally vary between different types of organisations. The measures described below are an example and are valid for a line-based road traffic company as described in the following paragraph.

One example of a company that performs line-based road traffic operates on a decentralised basis, meaning that the day-to-day operations are carried out by regional units responsible for both sales and production of the transports. The physical transports will be carried out either by the regional office's own vehicles or be subcontracted to smaller transporters. In addition, the head office has departments for finance, logistics and production planning, but these may be considered as support functions for the regional offices. In such an organisation the development of departmental quality measures will be confined to the regional offices, since these are the only departments that are involved in the production of the individual transports.

Departmental measures of shell quality components can also be developed for each regional office, since these handle all contacts with the respective customers.

Since, in this form of decentralised organisation, the regional offices operate in much the same way as a smaller, independent transport company, it is possible to establish a quality monitoring procedure that starts on a high level and then can be broken down into components. In this way, it is possible to monitor only a few quality components on a day-to-day basis. Only if the quality levels of those components are unsatisfactory it is necessary to perform a more detailed analysis of which function within the office is responsible for the deficiency.

Transports performed by the regional office itself can be monitored by measuring external and internal quality components as described in previous sections. Should these measures deviate from the desired levels it is necessary to exactly determine the cause behind the deviation. This process is illustrated by Figure 7.

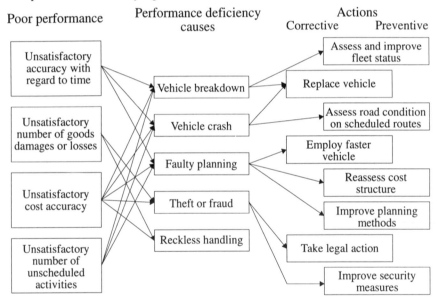

Figure 7. Method for determining the cause of core quality deficiencies in a regional office (example)

Based on the layout of Figure 7, events will take place in a left-to-right order. Initially, poor quality measurements will be gathered. This leads to the identification of the causes of poor quality performance. Following this, corrective actions to remedy the problem at hand are chosen. For future improvement, preventive actions are undertaken to ensure that the particular problem does not reappear.

In the case of transports performed by subcontracted transporters, the regional office may be considered as the customer of the subcontractors. Therefore, these transports need only be monitored with respect to the external quality components, i.e., accuracy and consignment conformance. Should a transporter consistently fail with respect to these measures the regional office must decide whether to assist the transporter or simply cease buying his services. Should it be considered essential to keep the transporter, the quality improvement process will follow the same procedure as the one used for quality improvement of the regional office's own transports.

Based on what has previously been stated about customer preferences, shell quality on the regional office level generally means to have short response time in customer communication. This will enable the regional office to take prompt corrective and preventive actions against customer complaints not discovered by the normal measuring activities. Alternative measures of response time are the time it takes the customer to get in touch with the right person in the transport company or the time it takes for a customer complaint to lead to corrective or preventive actions. Since the customer complaint after being received by the transport company should be treated as any other performance deviation, it is better to isolate the response time by simply measuring the time it takes the customer to get in touch with the right person. This measure can be stated in relative terms by comparing the actual response time with a predefined desired maximum response time. This measure would thus read

$$A_r = \left(1 - \frac{R_r - R_p}{R_p}\right)*100[\%], \ R_p \leq R_r; \ A_r = 100[\%], \ R_r \leq R_p, \text{ where}$$

A_r = Response time accuracy

R_r = Real response time

R_p = Projected response time.

Note that this measure would assume values over 100% if the formula on the left were to be valid for the absolute value of the difference between real response times shorter than the maximum allowed and the projected response. This would reflect that shorter response times have no negative effects, but would be mathematically problematic, since Rr=0 would yield Ar=200%. We therefore choose to let Ar=100% for all such cases.

Relations between measures and transport quality components

After having identified a set of quantitative measures, it is necessary to examine to what extent these measures cover the relevant transport quality dimensions of Jensen (1990) and logistics components of Heskett & Shapiro (1985) before employing the measures in the transport quality model described in a previous section.

The accuracy measure determines if a consignment is delivered to the right customer in the right place at the right time. Since an individual transport is examined, only the transport time dimension is covered; regularity and controllability can only be covered by studying a series of accuracy measures. The consignment conformance measure determines if a delivered consignment contains the right product in the right quantity and condition. Measured on an individual transport, this measure covers the goods comfort and transport security dimensions.

The transport cost accuracy measure determines if the transport has been performed at the right cost. This measure is not related to any of the transport quality dimensions. The activity accuracy measure does not relate directly to any logistics component. However, it covers the regularity and controllability dimensions, albeit on a more long-term scale than an individual transport, since these dimensions are related to the planning system and consistency of performances of the transport company.

The identified relations are summarised in Table 2. The symbols SQ1 through SQ5 in the table are used to uniformly denote the quality measures.

Table 2. Relations between transport quality measures, transport quality dimensions and logistics components

Measure	Type of quality component	Transport quality dimension	Logistics component
SQ_1 (Accuracy): $A_t = \left(1 - \frac{\|T_r - T_c\|}{T_c}\right)*100[\%]$	External	Transport time	Right place Right time Right customer
SQ_2 (Consignment conformance): $C_c = \left(1 - \frac{\|I_d - I_c\|}{I_c}\right)*100[\%]$	External	Goods comfort Transport security	Right product Right quantity Right condition
SQ_3 (Transport cost accuracy): $A_c = \left(1 - \frac{\|C_r - C_p\|}{C_p}\right)*100[\%]$	Internal	-	Right cost
SQ_4 (Activity accuracy): $A_a = \left(1 - \frac{A_u}{A_s}\right)*100[\%]$	Internal	Regularity Controllability	-
SQ_5 (Response time): $A_r = \left(1 - \frac{R_r - R_p}{R_p}\right)*100[\%]$	Departmental	-	-

Resulting measurement model

In the previous section, the links between measures and transport quality components were established. To determine the general quality level of a transport company and adapt the transport quality model for use in a transport company, however, it is also necessary to position the quantitative measures in the model. The internal components, i.e., cost accuracy and unscheduled activities, can be seen as reflections of the status of the production process and do not relate exclusively to either core or shell quality. The external measures, i.e., accuracy and consignment conformance, monitor the output of the production sub-processes and can hence be related to the production phase and the core quality concept. The response time measure monitors the administrative contacts with the customer and can hence be related to the consumption phase and the shell quality concept. The possible exceptions to this are transport legs at the beginning or the end of the transport chain that, since they involve direct contact with the transport customer, can be seen as relevant to the consumption phase as well.

Based on this, the transport quality model would assume the appearance of Figure 8. Note that the external components have been grouped in their entirety in the production phase, i.e., the shell quality-affecting nature of the first and last legs of the transport chain has been disregarded.

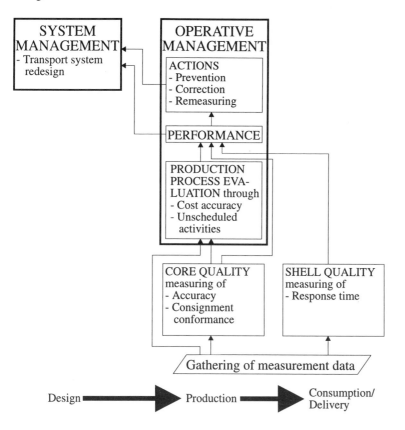

Figure 8. Model for measuring transport quality in line-based road traffic

All measures are made available to the operative management, both individually and in aggregated form. The aggregation of measures comprises attaching different weights to the measured components. These weights are chiefly determined by the customer emphasis on the different quality components. Based on the resulting performance data the operative management will take appropriate action. These actions are either in the form of corrective action or remeasuring within the operative management's sphere of control, i.e., the production and consumption phases, or in the form of preventive actions, which will prompt redesign of the transport system by the system management.

This model can be seen as a basic framework for monitoring and improving quality in a transport company. To be useable in practice it will probably need to be slightly modified to include local conditions and company preferences. However, the basic structure of the model will still be valid while the exact measures may be adapted accordingly.

Constructing a measure of the overall quality level

To facilitate employee awareness of the quality issue it is helpful to construct an overall measure of the quality level of the transport company. The model allows for this action since the measures contained in it are all expressed as percentages. They can therefore be aggregated into a percentage that would correspond to the relative accumulated deviation from the company's desired quality level. It can also be refined by using weight factors to the different sub-measures reflecting where the customers focus their attention as stated above. Using the sub-measure notation of Table 2, such an overall measure would read

$$Q = \frac{\sum_{i=1}^{5}\left(WF_i * SQ_i\right)}{\sum_{i=1}^{5} WF_i} * 100[\%] , \text{ where}$$

Q = Overall quality level [%]

WF_i = Weight factor i

SQ_i = Quality measure i.

TESTING THE MEASUREMENT MODEL

In order to test the measurement model, it is useful to compare it to quality monitoring systems currently employed in transport companies performing line-based road transports. To accomplish this, a case study was carried out at one such company. The purpose of this case study was to investigate what quality components were monitored and what kinds of measurement data gathering and evaluation methods were used.

Case study

The case study was performed as a review of the quality monitoring system of the company carried out in association with the person responsible for operations studies and evaluation at the central staff of the company. In this company the primary quality components that are monitored are as follows.

- Timeliness, i.e., the ability to pick up or deliver shipments on the same day as they enter or leave the district;

- Goods damages or losses, i.e., the number of consignments found to be damaged or missing;

- Availability, i.e., the degree of ease with which it is possible to make telephonic contact with company personnel;

- Invoicing errors, i.e., the number of invoices found to be incorrect;

- Signal list, i.e., the current amount of not corrected errors that are noticeable by the customers.

All these components, except the availability, are continuously monitored by the central staff of the company. The availability is measured by an independent firm that submits reports of the results twice yearly. In addition, the regional offices may, at their own discretion, monitor additional components. Results of the quality monitoring are distributed throughout the company twice a year.

The choice of components to monitor is made based on regular questionnaires that are sent to the customers of the company. The customers are asked to rate the importance of different components. This rating not only determines what components to monitor but also provides information that is used to calculate a general quality index. The quality index consists of the results for each of the monitored components weighted to reflect the rating of importance as stated by the customers in the questionnaire. The questionnaire is distributed about once a year.

The quality index is not used as an instrument for rating the performance of different regional offices. This is mainly due to the varying local conditions. For instance, some regional offices have difficulty getting high accuracy ratings because they cover large geographical areas. Instead, each regional office sets yearly performance targets in co-operation with the central staff.

Comparisons between measurement model and case study

The case study shows that there exist marked similarities between the measurement model and the quality monitoring system used in practice in the transport company. The timeliness component of the case study corresponds closely to the accuracy measure of the model. The main difference between these is the unit of measurement; timeliness is measured as the relative number of consignments that are delivered or picked up on the same day as they enter the area of the particular regional office whereas accuracy is measured as relative deviation from the planned arrival times at the respective customers. The goods damages or losses component of the case study is exactly the same as the consignment conformance measure of the model. The availability component of the case study is the same as the response time measure of the model, although they have differing data capture intervals. The signal list component of the case study can be loosely related to the unscheduled activities measure of the measurement model. The relation is that they both reflect the numbers of errors made by the transport company. The approaches taken to this are, however, very different, since the signal list focuses on errors as perceived by the customers whereas the unscheduled activities measure does not make this distinction. The invoicing errors component of the case study and the cost accuracy measure of the measurement model cannot be related to any other component or measure. The invoicing errors component monitors deficiencies that are unrelated to the physical transport and is therefore not included in the measurement model. The cost accuracy measure is very difficult to obtain for line-based transports, since it is hard to separate the cost of transporting an individual consignment that is transported together with various other consignments during different parts of the way. Obtaining this measure with any degree of precision would therefore require real-time information about the cargo of each vehicle at any given point of its route.

The relations between case study components and measurement model measures are shown in Table 3.

Table 3. Relations between the measures of the measurement model and the case study

Case study measure	Measurement model measure	Character of relation	Additional comments by company staff
Timeliness	Accuracy	Accuracy score normally worse than timeliness score	Seen as a useful marketing tool.
Goods damages or losses	Consignment conformance	Identical measures	Can only be measured at the terminals by the transport company.
Signal list	Unscheduled activities	Unscheduled activities score normally lower than signal list score	Can be used profitably to identify areas where operations can be improved. Aggregating it into the cost accuracy measure would be less informative.
Invoicing errors	-	-	-
-	Cost accuracy	-	Can currently only be measured manually on individual shipments.
Availability	Response time	Identical measures	-

CONCLUSIONS

The transport quality measurement model provides a framework for monitoring quality in a transport company performing line-based road traffic. It is also possible to create an overall measure of transport quality. Refining this into a concise definition of the term transport quality that goes beyond expressions such as 'conformance to specification' or 'fulfillment of customer requirements' is, however, more difficult. It can also be argued that such a definition is irrelevant, because of the varying requirements of different customers. Transport quality will hence mean different things for each transport company, although all must strive towards increased customer satisfaction. The transport company must therefore establish the quality requirements of its potential customer base to define transport quality in a company context.

The measurement model is largely supported by the case study of quality monitoring in a transport company. The main difference is that the model includes measures that, presumably because of data capture difficulties, are not present in the case study. The case study, on the other hand, includes measures that are not covered by the model. This later occurrence is, however, due to the demarcation of the measurement model. The conclusion is therefore that it is very well possible to implement the model in a company, provided that more accurate methods, primarily for capturing cost data, are developed. This involves finding data capture methods that are capable of gathering local information and then transmitting it to a central database for processing. This means that the transport industry often is at the front of information technology applications, such as bar-coding and satellite navigation systems.

One current trend is that transport companies offer complete solutions for realising the entire logistics chain of a customer. These undertakings are generally of a dedicated nature and designed to handle large goods flows. In such cases it is viable to determine the needs of each customer. The most direct way of doing this is of course to obtain the necessary information during the negotiations between the companies. However, it is also possible to anticipate the needs of potential customers by establishing the activity chain and its corresponding resource flows that is controlled by the customer. From this it is often possible to deduce the transport needs of the customer. One can therefore conclude that the measurement model is best suited for the regular services provided by the transport company, since dedicated logistics solutions normally require dedicated quality measures.

REFERENCES

Bolton, R.N., Drew, J.H. (1991). Longitudinal Analysis of the impact of service changes on customer attitudes, *Journal of Marketing*, Vol. 55, pp. 1-9.

Bowan, R., Muller, E.J. and Gordon, J. (1990). Shippers maintain quality standards, *Distribution*, Vol. 89, pp. 84-88.

Christopher, M., Schary, P. and Skjott-Larsen, T. (1979). *Customer Service and Distribution Strategy*, Associated Business Press, London.

Clow, K.E., Vorhies, D.W. (1993). Building a competitive advantage for service firms, *Journal of Services Marketing*, Vol. 7, pp. 22-32.

Gordon, J. (1989). The evolution of a quality campaign, *Distribution*, vol. 88, pp. 68-71.

Hellgren, J. (1994). *Model for Quantifying Transport Quality - Pilot Study*, Chalmers University of Technology, Göteborg.

Hopkins, S.A., Strasser, S., Hopkins, W.E. and Foster, J.R. (1993). Service quality gaps in the transportation industry: An empirical investigation, *Journal of Business Logistics*, Vol. 14, pp. 145-161.

Ishikawa, K. (1985). *What Is Total Quality Control?*, Prentice-Hall, Englewood Cliffs.

Jensen, A. (1990). *Combined Transport - Systems, Economics and Strategies*, Swedish Transport Research Board, Stockholm.

Lumsden, K. (1989). *Transportteknik*, Studentlitteratur, Lund. In Swedish.

McGinnis, M. (1990). The relative importance of cost and service in freight transportation choice: Before and after deregulation, *Transportation Journal*, Vol. 30, pp. 12-19.

Persson, G. and Virum, H. (red.) (1995). *Logistikk for Konkurransekraft*, Ad Notam Gyldendal, Oslo. In Norwegian.

Shapiro, R.D. and Heskett, J.L. (1985). *Logistics Strategy - Cases and Concepts*, West Publishing Co., St. Paul MN.

Strand, M. and Lumsden, K. (1989). *Produktkvalitet - Flödeskvalitet*, Chalmers University of Technology, Göteborg. In Swedish.

Wadsworth, H.M., Stephens, K.S. and Godfrey, A.B. (1986). *Modern Methods for Quality Control and Improvement*, John Wiley & Sons, New York.

Wilson, G.W. (1980). *Economic Analysis of Intercity Freight Transportation*, Indiana University Press, Bloomington.

Zeithaml, V.A., Parasuraman, A. and Berry, L.L. (1990). *Delivering Quality Service*, The Free Press, New York.

Zivan, S. (1987). Is quality under control?, *Distribution*, Vol. 86, pp. 100-101.

6

RELIABILITY IN SUPPLY-CHAINS AND THE USE OF INFORMATION TECHNOLOGY: A GEOGRAPHICAL VIEW

Anders Larsson

ABSTRACT

Modern industrial production is characterized by specialization and division of labour. One example is the subcontracting of parts of the production to outside firms which, combined with a growing use of time-compression strategies such as JIT, puts emphasis on the importance of information technology as a tool to control the production process.

Taking a geographical point of departure to these questions means to focus on how to use IT in order to communicate and organize industrial activities over large and small geographical areas. In the case of automotive suppliers reliability is presented as a strategic factor for the coordination of flexible time-compressed production systems. In order to meet the needs for reliability three different solutions are presented: relocation of suppliers, new transport arrangements or use of information technology.

Keywords: Subcontracting, time-compression, reliability, location, information,transportation

INTRODUCTION

Academic discussion about logistics and information technology is often defined within the borders of engineering or management. The aim of this chapter is to put forward the importance of geography in the analysis of problems concerned with the coordination of business activities and the use of IT.

Taking the point of departure in economic geography involves the basic assumption that organisation and coordination of economic activities in market economies to a great extent involves linking nodes of activity in space. In order to demonstrate the content of the rather abstract notion of "geographical view" there will be examples from the current restructuring process in the automotive supply industry.

The geographical dimension of coordination in automotive industry supply networks has become more important in the last decade because of a significant restructuring process, here represented by three major tendencies.

The first process is connected to the organisation of suppliers. Automotive assemblers want fewer and bigger suppliers to handle more complex products in order to minimise the assembly work in their plant, and to simplify coordination between themselves and the suppliers.

A second tendency to be considered here is the development of customer ordered production and time-compression strategies in the supply chains of automotive producers. One commonly cited example of this is the "just-in-time" production philosophy.

The last factor behind the need for a geographical view to logistics and IT is the globalization of business. Today there is a growing need for many companies to coordinate their activities in many countries and on several continents. The automotive supply industry provides an excellent example of this process, where the last years have seen tremendous restructuring activities towards large multinational automotive suppliers with activities in almost every automotive producing country.

These factors taken together create a situation where the coordination of activities in the supply network is becoming increasingly important and complex. Large multinational suppliers have to be mixed with small local firms, extreme sequential JIT-deliveries of some products are contrasted with several days of lead-time in other cases. Logistics does have a strategical role in managing these systems, and the use of information technology is a basic tool to ensure reliability and customer satisfaction.

GEOGRAPHICAL ASPECTS OF SUBCONTRACTING AND INDUSTRIAL RESTRUCTURING

Theories on industrial restructuring and location

In a very wide sense theories in economic geography can be seen as trying to explain spatial organisation of economic activity, a mission that of course can be done in a wide variety of ways.

Early theories on industrial location focused on how to find the least cost location of industries. Weber (1929) based his theory on the importance of transport cost to determine if location should be close to either the raw material source or to the market.

This cost minimizing and/or profit maximizing approach to location problems dominated up to the 1960's, see: (Smith 1981), when ideas based on behavioural theory challenged the traditional neo-classical thinking, see: (Pred 1967). The importance of decision making and the recognition of humans as working in an environment with non-perfect knowledge changed the question of location. Where a firm located its production could no longer be derived from a solely economic point of view because human irrational (in economic terms) behaviour had to be taken into account.

Structuralist geographers, see for example: (Massey 1984), argued that location of economic activity to a great extent was determined by structural factors outside the industrial organisation itself. They also shifted the focus of investigation towards questions of regional differences and social problems in society.

Much of the interest in industrial geography in the 1980's was focused on the multinational firm and how these firms became more and more internationalized and less dependent on nation-states (Dicken 1992). This development implied a more complex pattern of intra-firm relations, where large multinationals had to coordinate operations across continents. This "geography of the firm" concentrated much of its interest on the large multinationals as the driving force of economic development (Conti 1996). The geographical scale of investigation was preferably international and internationalization was mainly seen as a question of organization of production.

This is where information technology first appears as a key factor in economic geography because of its basic importance for the coordination of international business activities (Dicken 1992).

The last ten years have seen a couple of new and partly interrelated approaches to the analysis of business organisation and location. Yeung (1994) presents three recent approaches:

- *The post-Fordist/flexible specialization debate* with focus on how new flexible production systems replace the traditional Fordist mass production. Vertical disintegration and agglomeration economics leads to new industrial spaces and districts. This will be discussed in more detail later.

- *Regulation theory* puts emphasis on broader social structural changes in capitalism. It explains the breakdown of Fordism as a mismatch between the production technology and organisation (regime of accumulation) and the way society is organized and regulated (mode of regulation, see: (Aglietta 1976; Lipietz 1986; Tickell and Peck 1992).

- *The network approach* in economic geography has grown from an emerging realization that business organizations are changing rapidly. Traditional views of firm-relations as economic transactions between two actors are replaced by the image of networks where information becomes a strategic factor for competition. Relations are seen as interrelated and complex, see: (Axelsson and Easton 1992; Håkansson and Johansson 1992; Johansson and Mattsson 1988).

As a point of departure for the further discussion about supplier relations, location and the use of information technology we will look into the flexible specialization/industrial districts approach.

Flexible specialization and industrial districts

Questions of industrial reorganization with focus on a broader line of firm-environment theories have been addressed in the "post-Fordist/flexible specialization" debate (Asheim 1993; Schoenberger 1988a; Storper and Scott 1992). The ideas put forward about the development away from Fordist mass-production practices and social organization towards a flexible "post-Fordist" reality is based on the theories of the "regulationist-school".

In practice, this can be described as the fall of mass-production, both of the technology and the social organization that formed the ground for its development (e.g. wage-structures, consumption patterns, social security). What gradually is taking its place is a flexible form of production, where the use of modernities like computers and information technology allows manufacturing of traditional mass-production items (e.g. automobiles, electronics, white-ware) to be made in a customer-ordered fashion.

One of the fundamental ideas of the post-Fordist model of industrial restructuring is that consumer-preferences, competition and technology are changing in a way that forces firms to be more flexible in order to respond to quick market changes (Gertler 1988; Schoenberger 1988b).

This broad framework is the stating-point for the discussion on how flexible specialization leads to new geographical formations of industrial relations (Scott 1988). An industrial district according to (Scott 1992) is:

"... a localized network of producers bound together in a social division of labour, in necessary association with a local labour market".

The rationale behind spatial clustering is that close cooperation within a production process enables the district to have flexibility through small and medium sized firms working together and developing local competencies within certain segments of the economy.

One classical example is the tile industry in central Italy where districts are specializing on all the steps in the production process, including design, production, marketing and education. Other examples are taken from the electronics and film industry in California (Storper and Christopherson 1987).

This view of industrial restructuring is one of several schools of thought in the field and in contrast with the discussion about globalization and geographical dispersion of production systems.

In the following the flexible specialisation approach is used as a theoretical foundation to analyse the general shift from mass production to more flexible customer ordered products, especially in the automotive industry.

The question in focus is how firms manage their spatially dispersed or "global" networks and at the same time become more flexible and consumer-oriented. Here we will put forward three methods to overcome space: relocation, transportation and information technology. The discussion is focused on industrial supply chains, exemplified by the use of subcontracting linkages for input material in the production process in the automotive industry.

Division of labour and the use of subcontractors

A modern production process for mass-produced consumer goods, for example a passenger car or a dishwasher, contains a myriad of physical and information linkages. These linkages can be formally organized and institutionalized or they may be informal and depending on the mutual co-operation between individuals. Companies can choose to develop, manufacture and market products inside their own organizational domains or they may use outside sources to fill some of the requirements in the production chain. One method to organize these external relationships is to sub-contract work to independent outside firms.

Subcontracting is one of many methods to organize relationships between firms and can be manifested in physical and information links. But what makes subcontracting different from other forms of industrial relations? Holmes (1986) states that although the term subcontracting is ill-defined and used in many different ways, there is a general consensus that it describes a situation where one firm uses another independent firm to undertake production, processing or subassembling according to specification from the company offering the subcontract.

Subcontracting compared to any buyer-seller interaction involves a high degree of interdependence, the production-process of the contracting firm will be dependent upon the performance of the subcontractor while the subcontractor needs the buying firm for specifications, sometimes for raw-material and, most important, as a customer for the products made to specification. This means that the subcontractor in many cases cannot sell the product on the open market.

Sayer and Walker (1992) distinguish between relational contracting and subcontracting. Relational contracting, according to Sayer and Walker, are ongoing exchange relations and interactions between firms in order to create inter-firm connections without using market exchange; examples are: two-firm alliances, marketing agreements, licensing, and research alliances. Subcontracting is seen as a different form of relational contracting where there is involvement from one firm into another firm's process such as: product specifications, provision of materials and machines, technical and financial assistance and quality control.

TYPE OF	INTERFIRM RELATION	
LINKAGE	*Open-market relation*	*Subcontracting*
Product Organization Power	self-specified competitive independent	customer-specified relational (inter)dependent

Figure 1. Open-market and subcontracting outcomes of different types of linkages

So, what is the difference between purchasing input materials in a subcontracting relation and an ordinary open market situation? Figure 1 shows different aspects of inter-firm linkages and their outcome in subcontracting and open-market relations respectively. All three aspects of the buyer-seller relation indicate that subcontracting relations are characterized by closer co-operation and interdependence compared to open-market relations.

An industrial subcontracting arrangement can, according to the discussion above, be defined in terms of:

• Product specified by customer for use in his production process, which often results in specialized products not possible to sell on the open market. This means that the subcontractor needs a close and highly flexible relation with his customer in order to adapt to new specifications, and to avoid unmarketable stocks of old products.

• Relational and interdependent inter-firm linkages with a relatively high degree of communication and cooperation involved. There is of course a wide scale of power relations, ranging from marginal to total interdependence. The current trend in many industries is vertical disintegration and the transfer of more responsibility to subcontractors (Helper 1991; Imrie and Morris 1992).

Flexibility and industrial subcontracting

When a subcontracting relationship experiences a change from mass-production to customer-ordered flexible production there are several important organizational implications which can not be answered within one single theory of transition from Fordism to post-Fordism. However, some general trends can be seen, especially in the case of the physical production-process (Schoenberger 1988a):

• A significant reduction in cycle times in all parts of the production process, including R&D functions, production and sales. The idea of customer ordered production discussed above puts emphasis on the need for flexibility and close co-operation between subcontractor and customer. Otherwise a firm stands the risk to destroy a well functioning flexible intra-firm production system with inflexible subcontracting links.

• A need for greater sensitivity to market conditions, which is especially important when markets tend to be more segmented.

One set of methods, dedicated to improve product and process flexibility, is the "just-in-time" - production philosophy (Frazier, Spekman et al. 1988; Kalsaas 1995; Manoochehri 1984).

This set of ideas aims to reduce all unnecessary items in the production process by manufacturing just the right amount of products needed, just in time for its use in the process or for delivery to the customer. The practical use of JIT-methods is focused on time-rationalization in the entire process from raw-material to the end-user.

To make subcontracting relations function in this time-compressed environment, focus has to be on cooperation rather than competition, e.g. lead-firms give more responsibility to first tier subcontractors. Contacts are long-term and built on mutual trust, a condition for a successful implementation of the highly time-rationalized JIT-practices. The lack of time-limits or buffers in the production process has to be compensated for by stable long-term subcontracting links in order to supply the reliability needed, especially in physical relations (Imrie and Morris 1992; Ploos van Amstel 1990).

What we can see is that "time-compression" in a production process which is characterized by specialization and division of labour creates a demand for reliable coordination. In order to manage a complex flow of components with minimized time frames you need full control of the entire production process. This is where logistics services and the use of information technology will have an important impact (Germain and Dröge 1995; La Londe and Masters 1994).

Logistics - a key to coordinate flexible supply networks?

A geographical view of logistics focuses on the interaction between transport and industry, and how changes in the organization of production have spatial outcomes. This can for example be in the form of globalization of production, agglomeration of subcontractors, centralization of distribution centres or relocation of industrial activities out from urban areas.

The important starting-point is that transportation always involves movement of goods in time and space. This will have profound implications on customer expectations and the way transport producers can influence customer perceptions of the provided service. The development that has been laid out above focuses on how the transport/logistics customers are in a long-term development towards a situation where global and local systems of production are becoming equally important and have to be coordinated simultaneously.

One example is the Volvo Car Corporation in Europe. They have two assembly plants for passenger cars, one in Göteborg, Sweden and one in Gent, Belgium. The bulk of subcontractors are located in Europe, with some in Japan and the U.S. This geographical setting together with a clear strategy to reduce cycle-times and improve on flexibility has forced Volvo to rely heavily on their logistic capacity.

Just-in-time delivery several times a day is not possible unless the subcontractor is located close to the plant. But as relocation of a subcontracting firm is expensive and complicated, the Volvo logistics system has been forced to adapt to a combination of different conditions, from local subcontractors delivering every hour to Japanese suppliers of automatic gearboxes who deliver by ship every second week.

This exemplifies a situation where logistics as a strategy (Fabbe-Costes and Colin 1993) is a complex and multidimensional project held together by reliable information. Figure 2 exemplifies the difference between traditional standardised and flexible linkages between subcontractors and their parent firm.

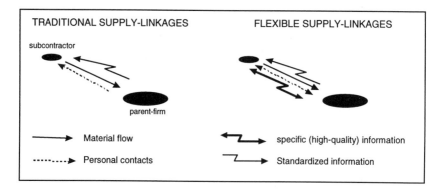

Figure 2. Traditional vs. new flexible supply linkages

Figure 2 shows in a schematic way how new flexible supply linkages have changed the demand for information. It is important to point out how information linkages are separated into two forms of communication: standardized information and specific, high-quality information.

Standardized information in this case is defined as every-day routine information that can be standardized and automated for high volume electronic communication. Examples are production schedules to suppliers or delivery instructions for JIT-deliveries.

Specific information is more common when the activities are difficult to plan in advance and with an unstructured form. This can be the case in R&D activities or in business negotiations where personal face-to-face contacts are vital.

In order to manage a flexible relationship there is a need for both standardized information and more specialized day-to-day interactions. In an ongoing study of the Volvo supply system there is clear evidence that both types of information are interrelated in a modern flexible supply linkage.

The next section will discuss the changing conditions for transportation and information in a new flexible supply system; special emphasis is put on the relation between reliability and information.

FLEXIBLE SUPPLIER NETWORKS AND RELIABILITY

Introduction

In the previous section we discussed industrial restructuring processes in terms of division of labour, globalization and flexibility. Subcontracting relations were used as a point of departure for a discussion about the growing need for coordination of production chains in an environment where throughput times are highly compressed. This development puts logistics into a strategical role as coordinator of flows of materials and information. This section is an attempt to discuss the importance of reliability in these types of supply relations.

Looking for quality-determinants in logistics services

As put forward earlier in this paper we can observe a tendency towards a division of labour and emergence of closer subcontracting relations in industry. This is of course a statement with many modifications depending on the actual industry and the context in which these linkages are situated.

If we look at this situation with "service-quality glasses" there is a fundamental difference in what different customers of logistics services will demand. If you have an industrial operation with several hundred subcontractors spread over a big area (e.g. Western Europe) that have to be coordinated with a highly time-rationalized assembly process in Sweden, there is need for integrated approaches. The logistics provider needs to integrate the subcontractor and the assembly plant with his own information system so that they can plan in advance and guarantee frequent and reliable deliveries.

This development can be compared to what Manrodt (1993) calls Service Response Logistics, with focus on the delivery of benefits that the consumer desires rather than on products. The importance of quality in such a concept is fundamental, and coincides to a great extent with the general definition of quality as the result of customer expectations and perceptions. One area where close cooperation in transport quality questions is of vital importance is between subcontractors and buyers in industrial production systems. The last decade has seen changes towards a closer relationship between purchaser and subcontractor, often described as "partnership" or "strategic alliances" (Brandes, 1990; Larsson, 1994).

Closer cooperation between firms in the supply chain puts more relative importance on communication, in information as well as goods. The emphasis on quality (products and processes), cost reduction (e.g. less inventory, "just-in-time"-delivery) and information/technology sharing (shortened lead-times) puts the overall logistics in a strategic position. It is most likely that a failure in providing the requested logistics service will have serious negative effects on the performance of the supply chain.

Gentry (1993) indicates that this development in industry is to be followed by strategic alliances between transport companies and subcontractors/buyers. Closer cooperation gives opportunities to meet new and more severe transport-quality demands in conformance with those guiding the entire production system. One effect of cooperation or alliances between transport companies and subcontractors/buyers that can be traced is the fact that the purchasing company tends to be more involved in decisions traditionally made by the carrier. In a study made among manufacturing firms in Minnesota and Wisconsin, Harper (1990) shows that the use of "just-in-time" production increased the cooperation between subcontractor and purchaser and at the same time increased the purchasing firm's control over the transport. The same tendency can be registered from a study of almost 700 American companies where almost 85 percent indicated involvement in selecting carriers for in-bound transportation (Gentry 1993).

A study of 120 subcontractors to two Swedish manufacturing firms gives similar indications (Larsson 1994). The companies were asked to specify who was responsible for planning and performance of transportation of components to the purchasing firms. Almost 90% of the answers indicated that the purchasing firm and/or the subcontractor were responsible for planning the transport, while the performance of the transport to a great extent (68%) was carried out by a transport company.

The examples above have been exclusively from the automotive industry and quality determinants may differ between firms and industries, but in general most writers agree that service quality is based on the interaction between customer expectations and perceptions. In order to develop any generalizations about service quality in the transport sector it is necessary to study logistics and quality in a broader framework. The service performed by logistics firms is an integral part of the total production process and quality aspects have to be analysed according to the performance of the entire system.

Reliability - a basic service-quality determinant in logistics?

In terms of service quality, quality of logistics services in a flexible subcontracting system has to be analysed as performance in relation to customer expectations, which emphasizes reliability as a central concept.

Reliability in the context of time-compressed flexible production can be defined as the capacity of the system to work according to production plan. The tighter the time-limits are the more important they are.

If we use the ideas of just-in-time (JIT) production as an example of time compression, there are a number of steps to be taken in order to manage subcontracting relations. Implementation of the JIT-concept should have the following strategical results (Kalsaas 1994; Manoochehri 1984):

- Significant reduction of inventory levels;

- Decreased lead-time requirements;

- Elimination of the production of unnecessary items;

- Production based on customer-orders in a "pull" -system;

- Flexibility in the production process (e.g. shortened set-up times, small batch production).

All of the items above have one common ingredient: they are focused on time rationalization. If a production-system is organized like this it is more or less dependent upon regular and frequent deliveries because the concentration on time-rationalization has reduced safety-margins in the supply chain to minimal levels. So, if we want both flexibility and time-compression in subcontracting relations, reliability in transportation and logistics will be the only way to ensure the stability needed to coordinate the production process.

This far in the paper we have discussed how supply systems have changed into international or global networks. At the same time we can see a development in many industries towards shortened lead-times and a more customer driven production process. From a geographical point of view we can identify a problem when a specialized international supply system meets the need for flexibility and low inventory levels.

In the following section there will be a discussion about the importance of information technology to uphold the need for reliability in industrial supply chains.

RELIABILITY AND LOCATION – THE ROLE OF INFORMATION IN FLEXIBLE SUPPLY CHAINS

Introduction

If we see reliability as a condition for flexible production systems to work, information becomes a strategic tool in order to coordinate the production chain. This is especially true for the relation between suppliers and assembly firms in the automotive industry. The use of new information and communication technology is often put forward as a solution to the complex task of coordinating such a modern flexible production system.

But it is important not to forget that the use of new and sophisticated technologies for communication is associated with physical linkages which may change radically as a result of new information environments. The geographical view used here takes its point of departure in the fact that information and location need to be studied simultaneously in order to better penetrate changes in the organization of supply networks in industry.

Reliability - spatial organization and information

As a point of departure for discussion a basic model is presented in Figure 3. Reliability is seen as being obtainable through either relocation, transportation or information. These factors are discussed in the following, but it is important to stress that in practice it is always a combination of factors that produces the final result.

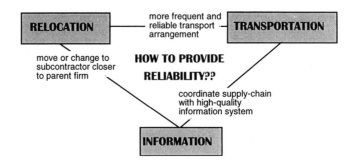

Figure 3. Relations between reliability, location, transportation and the use of information in time compressed supply chains

Figure 3 should be interpreted as a general model of the main concepts discussed in this paper. The implementation of time-rationalization strategies in subcontracting relations is closely linked to the need for reliability. With short cycle-times and virtually no inventory as backup, extreme reliability in the supply chain is needed to ensure smooth production.

RELOCATION

If the need for reliability toward the customers cannot be met, relocation of subcontractors is one option. This can be achieved either by trying to influence suppliers to move closer to the parent-firm, or by changing to local suppliers. The main advantages are:

- Shortened transport and overall lead-time;
- Better possibilities for late changes in orders;
- Less opportunities for errors during transport;
- Easier to control and therefore better reliability;
- Better conditions for personal face-to-face communication.

The extreme synchronized sequential high-volume "just-in-time" production of Toyota City in Japan is at one end of the spectrum. With the majority of the subcontractors within a 30 kilometre radius the geographical proximity provides the reliability needed for a flexible and yet mass-production process (Cusumano 1988; Mair 1992). But the Japanese example is in many ways unique, since it has been built as a greenfield site.

The situation is different in Western Europe where spatial clustering of automobile subcontractors cannot be found to the same extent as in Japan or even the U.S. (Sadler 1994). The European example is perhaps more interesting from a logistics point of view since the Japanese manufacturers (mostly located in the U.K.) use the traditional European subcontractor-base to a great extent. This situation involves, among other things, the coordination of supply chains from a number of countries where transport as well as cultural barriers influence reliability. Hudson (1995) discusses this in terms of "pseudo-just-in-time" with warehouses located close to the point of assembly to ensure reliability and regularity of deliveries.

One case of spatial proximity as a strategy in supply chains is Raufoss Automotive who supplies bumpers in sequential-synchronous JIT to the Volvo assembly-plant in Göteborg, Sweden. Ready-to-assemble bumpers are delivered hourly in sequence depending on model and colour. In order to match the need for reliability towards the customer the subcontractor has located a warehouse only 5 minutes transport time (4 kilometres) from the Volvo factory. This works as a "logistics point" with a 1-2 day buffer for the most frequent items, delivered from Raufoss plants in Sweden and in case of emergency from Belgium. Some assembly has been transferred to the warehouse, which gives flexibility because the final product can be changed just before delivery.

The production system is coordinated through a computerized on-line information system where the subcontractor gets delivery orders every seven minutes. These orders are printed out, the bumpers are collected, loaded on their own specially-built truck and delivered. For long term planning Raufoss also have access to data from Volvo.

In the case of Raufoss Automotive there are several aspects in favour for close location, especially when things don't work out the way they are expected to do. Face-to-face contact is fundamental in order to build a working relationship. Not only on managerial levels, many ad-hoc problems in the production were solved quickly and in an un-bureaucratic manner thanks to the short distance.

It has to be stressed that the Raufoss operations are more of a warehouse than an actual factory. This is a strategy used by a number of automotive suppliers in Europe (Munday, Morris et al. 1995), in order to match the demand for delivery frequency without moving their manufacturing. Instead you create a kind of "logistics-point" where the components are coordinated according to the production process of the parent-firm.

These ideas can be compared to so called "postponement-strategies" used in distribution where the final steps of the production (e.g labels, packages, manuals) are put as close in time to the final consumer as possible (Cooper 1994).

TRANSPORTATION

Improved transport arrangements in the form of increased frequency, speed or better infrastructure can have the same impact on reliability as relocation of subcontractors. It can bring the different parts of the supply-chain closer in time which leads to better control and flexibility of the entire production process.

Temporal proximity does not always imply spatial proximity. Geographical and infrastructural conditions may vary significantly between different contexts. One case is Japanese auto-manufacturers in the U.S. where JIT-production is maintained with a less localized subcontracting system than in Japan. In terms of reliability we can see how production in the North American context allows for subcontractors to be located 150 kilometres away from the assembly-plant while the corresponding distance in Japan is 40-50 kilometres (Mair 1992; Mair 1994).

Studies show that the introduction of time compression strategies into an already existing subcontracting network results in transport related activities such as increasing delivery frequency and reorganisation, rather than a significant geographical restructuring of suppliers (Harper and Goodner 1990; Helper 1991; Hill 1989; Larson 1991; Mair 1992; Reid 1995).

A study of Volvo and their Swedish subcontractors gives a similar result (Larsson 1994). Even though Volvo had a clear strategy towards customer driven production and time compression there were no signs of significant spatial reorganisation among subcontractors in the period studied, 1988-1992. Instead did most subcontractors indicate that the most extensive changes occurred in the transportation field. More than two thirds of the subcontractors answered that the frequency of delivery had increased. The major suppliers delivered at least daily compared to 1-2 times per week five years ago.

So, if we simplify the question of geography in supply chains, it seems to be transportation and not spatial reorganization that is the most common solution to create reliability in the production process. What we see is more frequent deliveries in smaller batches in order to match both time compression and inventory strategies. But this in turn creates substantially more traffic and consequently a bigger need for control and management. This leads to the third factor in our tentative model, how to use information to secure reliability in the supply chain.

INFORMATION

The introduction of new information and communication technology has had profound implications on the management of supply chains in the last ten years. Giannopoulos (1993) defines three levels of road freight operations where information technology is used: i) management and logistics functions where IT is used mainly for communication and long-term planning, ii) fleet management functions such as route planning and scheduling, iii) vehicle management including trip planning and cargo identification.

The discussion in this paper will concentrate on the first level. Focus is on the management of actors in a supply chain and the impact of new information technology as a means of communication. It is important to stress that *the communication aspect* of information technology is in the center of interest in this paper.

A study of two large Swedish corporations (Lorentzon 1995) and their use of information and communication technology uses the following means of communication : telephone, telefax, telex, e-mail, EDI, mail, business trips and video-conference. One of the conclusions drawn from the study is that the use of IT is different depending on corporate functions. The supply chain is one of the most developed areas where firms use electronic data interchange (EDI) for large volumes of standardized information, while functions with more unstructured information tend to use the more traditional phone and fax together with face-to-face contact.

The case of Raufoss Plastal Automotive discussed earlier is an example of how information in the supply chain is segmented into structured and un-structured forms. By using standardized forms of communication it is possible for Volvo to provide Raufoss with bar-coded orders every seven minutes. At the same time there are still a lot of phone, fax and personal contacts between the parties regarding unstructured information such as product or process problems.

Germain and Dröge (1995) show in a study of U.S. manufacturers that the implementation of just-in-time strategies is positively related to the use of EDI. The authors conclude that the use of EDI has more to do with the extent of JIT-implementation rather than type of industry, technology or business environment.

The development of time based strategies in manufacturing is closely related to the handling of information in the supply chain. Daugherty and Ellinger (1995) indicate in a study of warehouse-firms in the U.S. that there is a crucial link between customer responsiveness and information availability.

Although the IT-development in many ways is linked to technology, it is important to point out that information as such not necessarily has to be communicated through computer networks, fax machines or mobile phones. And as the examples above indicate there is a difference between forms of information. And the use of time as a competitive strategy has to go hand-in-hand with an information strategy recognizing the difference between standardized large volume information and unstructured ad-hoc information.

The discussion so far has been about manufacturing firms where information, especially standardized high-volume, in most cases is connected to a physical flow. This means that information technology cannot more than partially influence the location of subcontractors.

The situation is different in the case of supply chains in certain service industries where the product might be a produced service such as information, a bank transaction or a holiday trip. One example is the finance sector, which thanks to global networks and computers does have 24 hour instant access to information around the globe. Geographically you can be located anywhere as long as you have the relevant technology to gain access to information.

The point to be made here is that information technology is a vital tool for firms that use time as a competitive strategy, but as long as there is a corresponding physical flow along the supply chain we must not forget about the material limitations to our often specialized and spatially dispersed production systems.

CONCLUSION

Looking at questions of information technology from an economic-geography point of departure emphasizes problems of spatial coordination of business activities. This means that information technology is defined in terms of its communicative function, rather than technological or economic performance.

The discussion in this chapter has been centered around questions of reliability in supply networks, especially in the automotive industry. Reliability was put forward as a strategic factor in the coordination of flexible time-compressed production systems.

In order to meet the needs for reliability three different solutions were presented: relocation of suppliers, new transport arrangements or use of information technology. These three factors are of course working in a dynamic interplay where each has more or less importance in different contexts.

One general conclusion to be drawn from the economic geograpic literature is that new information technology is one of the main forces behind the growth of global firms (Dicken 1992). The possibility to coordinate activities across the globe through information technology has widened the scope of supply networks. In the automotive industry we can see a clear change from national to international or global sourcing and an intensified competition and restructuring.

The relative importance of information has increased and this has implications on the location and transportation aspects of suppliers. Changes can be seen in the automotive industry where the increasing move towards customer ordered production and shortened lead-times propelled by information technology has created a segmented supply system. Components with a large number of variants and/or colours are often delivered in synchronous sequence several times per day, while more standard items are once or twice per week.

In terms of location changes in such a system we have a complex situation with different conditions on different "levels" in the supply network. The general conclusion from the literature and from an ongoing PhD-project by the author is that a high delivery frequency often corresponds with an extensive use of information, and in a growing number of cases, also with proximity in location.

This does not necessarily mean that suppliers in the automotive industry move their operations closer to their main customer. In many cases there is a situation where a new facility is opened just to function as a warehouse for delivery purposes while the actual production remains in its original location.

So, together with a more complex system of suppliers we are also facing a situation where single supplier firms or constellations have very different conditions within the same organization. Being an automotive supplier can no longer be regarded as a simple one-dimensional function but rather a complex enterprise where reliability conditions vary and so does the use of information, transportation and geographical location.

ACKNOWLEDGEMENT

This paper is a part of a PhD-project financed by the Swedish Transport and Communications Research Board (Grant No. 89-139-43).

REFERENCES

Aglietta, M. (1976). *A theory of capitalist regulation.* London: New Left Books.

Asheim, B. T. (1993). Industrial districts, inter-firm co-operation and endogenous technological development: the experience of developed countries. Geneva: UNCTAD.

Axelsson, B. and G. Easton, Eds. (1992). *Industrial Networks: A New View of Reality.* London: Routledge.

Brandes, H. and J. Lilliecreutz (1990). *Structural Changes in Networks*. Paper presented at the 6th I.M.P Conference: "Research Developments in International Marketing and Purchasing". Milan, Italy

Conti, S. (1996). "Four Paradigms of the Enterprise System" In: S. Conti, E. J. Malecki et al, (eds.). *The Industrial Enterprise and Its Environment: Spatial Perspectives* London: Avebury.

Cooper, J., M. Peter, et al. (1995). *Supply Chain Dynamics and the Environment - a study of six industry sectors*. Cranfield University, Centre for Logistics & Transportation. CCLT Report No. 1.

Cooper, J. (1994). "The Global Logistics Challenge" In: J. Cooper, (eds.). *Logistics and Distribution Planning: Strategies for Management* London: Kogan Page.

Cusumano, M. A. (1988). "Manufacturing Innovation: Lessons from the Japanese Auto Industry." *Sloan Management Review* 30(1): pp.29-39.

Daugherty, P. and A. Ellinger (1995). "Information accessibility: customer responsiveness and enhanced performance." *International Journal of Physical Distribution and Logistics Management* 25(1): pp. 4-17.

Dicken, P. (1992). *Global Shift*. London: Paul Chapman Publishing.

Fabbe-Costes, N. and J. Colin (1993). "Formulating Logistics Strategy" In: J. Cooper, (eds.). *Logistics and Distribution Planning: Strategies for Management* London: Kogan Page. Pp. 12-23.

Frazier, G., R. Spekman, et al. (1988). "Just-In-Time Exchange Relationships in Industrial Markets." *Journal of Marketing* 55(October 1988): pp. 52-67.

Gentry, J. (1993). "Strategic Alliances in Purchasing: Transportation Is the Vital Link." *International Journal of Purchasing and Materials Management* 29(3): pp. 11-17.

Germain, R. and C. Drö ge (1995). "Just-in-time in context: predictors of electronic data interchange technology adaption." *International Journal of Physical Distribution and Logistics Management* 25(1): pp. 18-33.

Gertler, M. S. (1988). "The Limits to Flexibility: Comments on the Post-Fordist Vision of Production and Its Geography." *Transactions of the Institute of British Geographers* 13: pp. 419-432.

Giannopoulos, G. (1993). "Information technology innovation in road freight transport" In: G. Giannopoulos and A. Gillespie, (eds.). *Transport and Communications Innovation in Europe* London: Belhaven.

Harper, D. V. and K. S. Goodner (1990). "Just-in-Time and Inbound Transportation." *Transportation Journal* 30(2): pp. 22-31.

Helper, S. (1991). "How Much Has Really Changed between U.S. Automakers and Their Suppliers?" *Sloan Management Review* (Summer 1991): pp. 15-28.

Hill, R. C. (1989). "Comparing transnational production systems: the automobile industry in the USA and Japan." *International Journal of Urban and Regional Research* 13(3): pp. 462-479.

Holmes, J. (1986). "The Organization and Locational Stucture of Production Subcontracting" In: A. J. Scott and M. Storper, (eds.). *Production, Work and Territory: The Geographical Anatomy of Industrial Capitalism.* London: Allen & Unwin.

Hudson, R. (1995). *The end of mass production, the end of the mass collective worker and their respective geographies? Or more old wine in new bottles?* Paper presented at the 1995 annual conference of the Institue of British Geographers, 3-6 January. University of Northumbria, Newcastle, UK

Hå kansson, H. and J. Johansson (1992). "A model of industrial networks" In: B. Axelsson and G. Easton, (eds.). *Industrial Networks: A New View of Reality.* London: Routhledge.

Imrie, R. and J. Morris (1992). "A Review of Recent Changes in Buyer–Supplier Relations." *Omega* 20(5/6): pp. 641-652.

Johansson, J. and L.-G. Mattsson (1988). "Internationalisation in Industrial Systems – A Network Approach" In: N. Hood and J.-E. Vahlne, (eds.). *Strategies in Global Competition.* London: Croom Helm.

Kalsaas, B.-T. (1995). *Transport in industry and locational implications: "just-in-time" principles in manufacturing, generation of transport and the relative impact on location. Scandinavian and Japanes experiences.* Doktor ingeniø ravhandling 1995:118, Trondheim: Norges Tekniske Hø gskole, Institutt for by- og regionplanlegging.

Kalsaas, B. T. (1994). *Just-in-Time Delivery in the Automotive Industry and the Location of Subcontractors.* Paper presented at the 34th European Congress of the Regional Science Association. Groningen, The Netherlands.

La Londe, B. and J. Masters (1994). "Logistics Strategies for the USA" In: J. Cooper, (eds.). *Logistics and Distribution Planning: Strategies for Management* London: Kogan Page.

Larson, P. D. (1991). "Transportation Deregulation, JIT, and Inventory Levels." *Logistics and Transportation Review* 27(2): pp. 99-112.

Larsson, A. (1994). *Changing Subcontracting Linkages: Geographical Aspects of Just-in-Time manufacturing in Two Swedish Firms.* Occasional Papers 1994:9, Gö teborg: Department of Human and Economic Geography, School of Economics, University of Gö teborg.

Lipietz, A. (1986). "New Tendencies in the International Division of Labor: Regimes of Accumulation and Modes of Regulation" In: A. J. Scott and M. Storper, (eds.). *Production, Work, Territory: The Geographical Anatomy of Industrial Capitalism.* Boston: Allen & Unwin.

Lorentzon, S. (1995). "The use of ICT in TNCs: A Swedish perspective on the location of corporate functions." *Regional Studies* 29(7): pp. 673-685.

Mair, A. (1992). "Just-in-Time Manufacturing and the Spatial Stucture of the Automobile Industry: Lessons from Japan." *Tijdschrift voor economische en sociale geogafie* 83(2): pp. 82-92.

Mair, A. (1994). *Honda's Global Local Corporation.* New York: St .Martin's Press.

Manoochehri, G. H. (1984). "Suppliers and the Just-in-Time Concept." *Journal of Purchasing and Materials Management.* Winter: pp. 16-21.

Manrodt, K. B. and F. Davis Jr (1993). "The Evolution to Service Response Logistics." *International Journal of Physical Distribution & Logistics Management* 23(5): 56-64.

Massey, D. (1984). *Spatial Divisions of Labour: Social Structures and the Geography of Production.* London: Macmillan.

Munday, M., J. Morris, et al. (1995). "Factories or Warehouses? A Welsh Perspective on Japanese Transplant Manufacturing." *Regional Studies* 29(1): pp. 1-17.

Ploos van Amstel, M. J. (1990). "Managing the Pipeline Effectively." *Journal of Business Logistics* 11(2): pp. 1-25.

Pred, A. (1967). *Behaviour and Location: Foundations for a Geographic and Dynamic Location Theory - part I and II.* Serie B Human Geography No. 27, Lund: Lunds Universitet, Geografiska institutionen.

Reid, N. (1995). "Just-in-time Inventory Control and the Economic Integration of Japanese-owned Manufacturing Plants with the County, State and National Economics of the United States." *Regional Studies* 29(4): pp. 345-355.

Sadler, D. (1994). "The Geographies of Just-in-Time: Japanese Investment and the Automotive Components Industry in Western Europe." *Economic Geography* 70(1): pp. 41-59.

Sayer, A. and R. Walker (1992). *The New Social Economy: Reworking the Division of Labour.* Cambrige, MA & Oxford: Blackwell.

Schoenberger, E. (1988a). "From Fordism to Flexible Accumulation: technology, competititve strategies and international location." *Environment & Planning D: Society and Space* 6: pp. 245-262.

Schoenberger, E. (1988b). "Thinking About Flexibility: A Response to Gertler." *Transactions of the Institute of British Geographers* 14: pp. 98-108.

Scott, A. J. (1988). *New Industrial Spaces.* London: Pion Ltd.

Scott, A. J. (1992). "The role of large producers in industrial districts: a case study of the high technology system houses in southern California." *Regional Studies* 26: pp. 265-75.

Smith, D. M. (1981). *Industrial Location: An Economic Geographical Analysis. Second Edition.* New York: Wiley & Sons.

Stalk Jr., G. and T. Hout (1990). *Competing Against Time.* New York: The Free Press.

Storper, M. and S. Christopherson (1987). "Flexible Specialization and Regional Industrial Agglomerations: The Case of the U.S. Motion Picture Industry." *Annals of the Association of American Geographers* 77(1): pp. 104-117.

Storper, M. and A. J. Scott, Eds. (1992). *Patways to Industrialization and Regional Development.* London: Routledge.

Tarski, I. (1987). *The Time Factor in Transportation Processes.* Developments in Civil Engeneering 15, Amsterdam: Elsevier.

Tickell, A. and J. A. Peck (1992). "Accumulation, regulation and the geographies of post-Fordism: missing links in regulationist research." *Progress in Human Geography* 16(2): pp. 190-218.

Weber, A. (1929). *Theory of the Location of Industries*. Chicago: University of Chicago Press.

Yeung, H. W. C. (1994). "Critical reviews of geographical perspectives on business organizations: towards a network approach." *Progress in Human Geography* 18(4): pp. 460-490.

7

A FRAMEWORK EFFICIENCY MODEL FOR GOODS TRANSPORTATION: WITH AN APPLICATION TO REGIONAL BREAKBULK DISTRIBUTION

Anders Samuelsson
Bernhard Tilanus

ABSTRACT

A general framework efficiency model for goods transportation is proposed. It is specified for the case of regional breakbulk distribution. Quantitative data on the magnitude of the partial efficiencies, their dispersion, and their potential for improvement have been collected from experts in the field. What should be done next is select promising partial efficiencies, analyse them statistically, find causal relations and recommend improvements.

Keywords: Efficiency, performance, measurement, transportation, distribution

INTRODUCTION

In the recession of the early nineties, interest in the efficiency of operations has increased. In the Dutch research project LOPER (Ruijgrok et al., 1992), it was tried to assess the logistical performance of logistical service organizations. Also, the efficiency of transportation was assessed (Tilanus, 1990).

Along similar lines, we are investigating the efficiency of goods transportation. But applying the French principle of 'reculer pour mieux sauter', we take a long take-off run. Our starting point is a theoretical, ideal transportation situation. Successively applying a number of efficiencies in various dimensions, each of which is broken down in further partial efficiencies, we finally arrive at the actual transportation situation. The overall efficiency of the actual transportation situation is assessed as the continuous product of all defined partial efficiencies.

Efficiencies may be assessed ex post, and analyzed for better understanding the transportation process. If the causal relations between efficiency and explanatory factors are understood, a potential for improvement may be assessed. A low efficiency may be combined with a high, or a low, potential for improvement; likewise, an already high efficiency may still have a high, or a low, potential for improvement. If potentials for improvement are known, further analysis and recommendations for improvement may be focussed on those partial efficiencies that reveal a high potential for improvement.

Efficiencies may be assessed ex ante as well, and used for planning purposes. If they are used for planning purposes, they must be forecast. Even if the simple average efficiency of the past is extrapolated into the future, this is a forecast. Efficiencies should then be considered as stochastic variables with a mathematical expectation and a variance. For planning purposes, the dispersion of an efficiency plays a role. Unstable efficiencies jeopardize exact planning. Margins should be incorporated. Reducing the dispersion of an efficiency may be an important goal for improvement!

One may take either a Bayesian, or a non-Bayesian view. In the Bayesian view, there is prior (subjective) knowledge, there is a forecast and/or a plan. An appropriate measure of dispersion would then be the mean square deviation from the forecasted value (standard error). In the non-Bayesian view, there is no prior knowledge or subjective probability. An appropriate measure of dispersion would then be the mean square deviation from the average value (standard deviation). We did not make this distinction in our interviews with experts from the field, and they did not raise the problem to us. However, the referees pointed out the problem to us.

Furthermore, transport efficiency, as to be discussed in this article, is only a subset of supply chain efficiency. The supply chain does not only consist of transformation of place, i.e., transportation, but also of other transformations, e.g., transformation of time (storage) or form (assembly, value added logistics). For these other transformations, useful starting points for overall efficiency measures will have to be developed in the future. This is outside the scope of this article.

Supply chain efficiency, in turn, should be seen in a still wider context of efficiency objectives. Different agencies/organizations will have efficiency objectives which may be in conflict with one another. For example, government may focus on securing reductions in pollution levels or energy consumption, whereas companies may want to make use of transport as a result of rationalization of depot locations made possible by improved infrastructure (Quarmby, 1989).

Even when confining our efficiency objectives to the scope of the individual transportation firm, we realize that measuring physical efficiencies is only one step on the road to the ultimate goal of the firm (Chow, Heaver and Henriksson, 1994; Clarke, 1991). Research may focus on other physical efficiencies than the ones dealt with in this article, e.g., transportation lead time (Ploos van Amstel, 1990), or the use of energy (Walker and Wirl, 1993; McKinnon, Stirling and Kirkhope, 1993). After physical efficiencies, come cost efficiencies. Next to efficiency ('doing things right'), effectiveness ('doing the right things') should be investigated (Gattorna, 1988; Mentzer and Konrad, 1991). Customer service and customer satisfaction may be intermediate goals to maximize revenues (Heskett, 1994; Livingstone, 1992; Rhea and Shrock, 1987). Total quality management may be the firm's professed objective (Foster, 1989; Novack, 1989). Revenues minus costs give profits, which to maximize may be the firm's objective. Again, profits may be maximized in the short, or in the long run. Finally, the firm may have the multicriteria goals of cherishing the customers, the shareholders and the personnel, with the ultimate objective of survival.

The structure of this article is as follows. In the next section, we present the general framework efficiency model for goods transportation. Then a specific efficiency model is developed for the regional truck distribution/collection link of international goods transportation by any mode. Along with it, expert opinion is presented on the magnitude of the efficiency, the dispersion, and the potential for improvement. In the final section, conclusions and recommendations for further research are given.

A FRAMEWORK EFFICIENCY MODEL FOR GOODS TRANSPORTATION

Efficiencies

Efficiencies are defined as fractions, or percentages. An efficiency of 100 per cent means that the theoretical, ideal situation is attained. An effiency above 100 per cent means that the theoretical, ideal situation is surpassed. This can only be the case incidentally, for instance if a speed limit is 80 km/h, but the actual driving speed is above that; or if a truck loading capacity is 40 m3, but actually more is loaded because the driver takes some extra boxes into his cabin.

An efficiency below 100 per cent means that the actual situation falls short of the theoretical, ideal situation. Then it depends on the dimension of measurement, if the actual value should be in the nominator and the theoretical, ideal value in the denominator, or the other way round. If the dimension measured is 'good', e.g., capacity, time or profit, then the actual value should be in the nominator. If the dimension measured is 'bad', e.g., distance or costs, then the actual value should be in the denominator.

If the optimum value of the dimension measured is neither the maximum, nor the minimum, but something in between, defining an efficiency measure is more problematic. One might put the actual value in the nominator if it is below the optimum value, and in the denominator if it is above the optimum. Or else one might drop the dimension altogether and seek another dimension in which a maximum or a minimum value is desirable.

An example may be speed. In road transportation, the optimum speed may be the maximum speed allowed by the speed limits valid on a given kind of road for a given kind of vehicle. But if there are no speed limits, like on the German Autobahn, we are in trouble trying to define a speed efficiency. Likewise in shipping, since there are no speed limits imposed on the ocean, we cannot define a maximum speed as the theoretical, ideal situation, but we must define an optimum speed. Here, we might switch to another dimension, like energy or costs, and define an efficiency as minimum energy or minimum costs divided by actual energy used or actual costs incurred. For the time being, we will avoid problems with non-minimum or non-maximum optima, and assume that for each efficiency, the variable concerned should be maximized or minimized.

General physical efficiencies of goods transportation

The starting point for assessing the efficiency of goods transportation, in any actual, past or future, situation for any transport mode in any link of the transportation chain, is a theoretical, ideal situation. For goods transportation, a theoretical, ideal situation would be non-stop, full-time transportation from origin (A) to destination (B), and back, along a minimum distance, at a maximum speed, of the net, pure product, at full capacity of the means of transportation. Means of transportation or vehicle stands for any transport modality: truck, train, barge, coaster, deep sea ship, airplane or pipeline. The theoretical, ideal situation stands for the (unattainable) maximum transportation output.

The justification for choosing the non-plus-ultra of transportation as a starting point for our efficiency model is that we are dealing in this article with the physical efficiency of transportation, which is, admittedly, only part of the efficiency of a supply chain, which is part of the final objective of the firms concerned, which is part of the ultimate goal of society.

Overall efficiency (E) in the general goods transportation situation tells one what the actual transportation output is as a percentage of the theoretical, maximum output. Overall efficiency in the general goods transportation situation consists of the continuous product of four dimensional efficiencies with respect to time, distance, speed and capacity:

$$E = T * D * S * C. \qquad (1)$$

The dimensional efficiencies with respect to time, distance, speed and capacity are explained below.

Time efficiency (T) is the percentage of time the means of transportation is actually transporting goods. The theoretical, ideal situation is literally non-stop, 24-hours a day. Many factors, each representing a partial efficiency, detract from this theoretical maximum. The vehicle may only be operated in certain shifts, it may need maintenance, it must stop for loading and unloading, to name a few. Which partial efficiencies it is useful to distinguish, should be considered in every specific situation.

Distance efficiency (D) is the percentage by which maximum transportation output is reduced by not using the shortest route between origin and destination. Usually the ideal, shortest route will be the minimum, radial distance. This is detracted from by factors like infrastructure, groupage of goods, etc. Distance efficiency is then the radial distance divided by the actual distance.

Speed efficiency (S) is the percentage by which maximum transportation output is reduced by not transporting at maximum speed. We may also consider optimum output in an economic sense. Then, optimum transportation output would be equivalent to transporting a given amount of goods between A and B in the cheapest way. We may also define efficiency in terms of energy consumption or even environmental pollution. If all external costs (energy consumption, environmental pollution) are internalized, economic efficiency is equivalent to energy or environmental efficiency. Still, a higher than optimal speed may decrease economic efficiency, whereas physical efficiency is increasing linearly with speed by definition. In practice, however, for many modes of transport and links in the transport chain, the optimum speed is the same as the maximum allowable speed, defined by technology (trains) or law (trucks).

Capacity efficiency (C) refers to the capacity of the vehicle and is measured in terms of whichever capacity is binding: weight or space. It is defined as the percentage by which maximum transportation output is reduced by not transporting at optimum capacity. Similar considerations as with speed should be made. In practice, optimum capacity may be the same as maximum capacity, either in terms of weight or in terms of size of the vehicle, as determined by technology (trains, barges) or law (barges, trucks). For seaships, airplanes and pipelines, maximum physical capacity is so large as to be difficult to assess. A solution will have to be found depending on the situation (modality and link in the chain).

This concludes our general framework efficiency model for goods transportation. The description may have seemed quite vague and abstract. But it allows, we hope, for a more concrete specification in each transportation mode/link situation. The way the dimensional efficiencies are further broken down into partial efficiencies will warrant pursuing this continuous product model further. For a specific example, let us consider the case of regional breakbulk goods distribution/collection by road.

APPLICATION TO REGIONAL BREAKBULK DISTRIBUTION/COLLECTION BY ROAD

The international, or long-distance, transportation chain of breakbulk goods consists of three links: in one country, or region, the goods are collected in a roundtrip by a truck; they are regrouped to batches suitable for the long-distance vehicle, which may be a large-volume or a heavy-weight truck, or a shuttle-train, or an inland or a Rhine barge, or a shortsea ship, and transported in the vehicle to another country, or region; there, they are regrouped into 'tours' for an area within the other country, or region, and distributed in a roundtrip by a truck; this truck may in the same roundtrip collect goods for transportation from this country, or region, to other countries, or regions. A number of such chains may be integrated into a hub-and-spoke transportation/distribution network (Irestähl, 1984; Taha and Taylor, 1994).

In Europe, even though the economic borders are eliminated, there still exist countries with huge psychological, linguistical and cultural walls around them. Distribution and collection of breakbulk goods is usually done by country. If there were no countries, like in the USA, it would be done by state or by region. We shall use the term regional distribution/collection and consider only this link in international, or long-distance, transportation. Regional distribution/collection of breakbulk by road thus consists of trips usually for the duration of one day, from or to one depot. The number of deliveries per daytrip is something between 15 and 50.

APPROACH

For all partial efficiencies to be introduced into the model, we have sought expert opinion as to their usefulness, their order of magnitude, their stability and their potential for improvement. We will communicate this information as we discuss the partial efficiencies.

The approach may be called semi-Delphi:

1. We have first broken down the general efficiencies into partial efficiencies as we saw fit, and proposed values for them.

2. We then interviewed 15 experts from road haulage and forwarding companies in Sweden and the Netherlands, asking their comments as to the breakdown and their quantitative assessments (see Acknowledgement).

3. We incorporated the comments and averaged the quantitative assessments, returned the results to the experts and asked their second round opinion, stressing the fact that circumstances and quantitative values in different companies may be different. This was followed up by telephone interviews.

4. After the second round, we decided on the breakdown and quantitative assessment given in this article. It should be noted that almost consensus was reached about the usefulness of the breakdown into individual partial efficiencies and about the estimated values assigned to them.

Partial time efficiencies

The point of departure for time efficiency (T) in regional breakbulk distribution is a road vehicle, transporting full-time, non-stop, goods from A to B, and back.

We shall break down time efficiency into four partial efficiencies, called business time (TB), availability (TA), utilization (TU) and driving factor (TD):

$$T = TB * TA * TU * TD. \quad (2)$$

Business time efficiency (TB) is the number of hours per annum that business is going on, divided by 8760. If a professional carrier business is only closed between Saturday 16.00 hrs and Monday 4.00 hrs each week the year round, its TB is 0.79. Business time is mostly constrained by the opening hours of the shipments consignees businesses! Business time efficiency may significantly be improved by night distribution. Experts assess this efficiency at 0.37, stable, but with potential for improvement, e.g., by night distribution.

Availability (TA) is the percentage of business time that the vehicle is available for use. An optimum balance should be found between preventive and corrective maintenance. Breakdowns, which may be caused by accidents, should be avoided. Repairs, either in one's own workshop or in an outside workshop, should not be delayed by lack of spare parts. This efficiency is generally well attended to by management. It is unstable for mostly external reasons and hard to improve. It is assessed at 0.94.

Utilization (TU) is the percentage of time available that the vehicle is actually used. Non-use may result from lack of business. It is the commercial department's job to achieve a high utilization, but this should be balanced by a degree of flexibility. Business administrators often calculate unit costs on the basis of a 'normal' degree of utilization of 0.80. Because our definition is slightly different, time used divided by time available, TU is put at 0.91. This is a fluctuating efficiency, depending on business circumstances. Commercial departments have quite a number of possibilities to increase the average and reduce the fluctuations.

Driving factor (TD) is the percentage of time when the vehicle is utilized that the vehicle is actually driving and transporting goods. The losses of driving time are due to stops for (a) resting time for the driver prescribed by law, (b) extra, unnecessary stops by the driver, (c) loading/unloading time at the depot or transshipment point, and (d) loading/unloading time at the addresses visited. Loading/unloading time here is a euphemism for the total time that a call takes, which may consist of long waiting times, administrative delays, etc. The smaller the shipments are in a distribution trip, the more addresses must be visited, the larger the total stop time and the lower the driving factor becomes. In the distribution of small packages, stop times of over 50 per cent occur. In a typical breakbulk distribution situation, total stop time may be 30 per cent, in other words, the driving factor will be 70 per cent. The experts' assessment was 50 per cent. The driving factor is of major concern for transportation management. It is not only low on average, also its fluctuations are a cause of disruption of the distribution process. This aspect of transportation management is sometimes called stop management.

For the numerical assessment given here, time efficiency would be

$$T = 0.37 * 0.94 * 0.91 * 0.50 = 0.16,$$

with possibilities for improvement especially of business time (0.37) and driving time(0.50).

Partial distance efficiencies

The point of departure for distance efficiency (D) is radial distance between A and B.

We shall break down distance efficiency into five partial efficiencies, called infrastructure factor (DI), backhaul factor (DB), routing factor (DR), detour factor (DD), and actual trip execution efficiency (DA):

$$D = DI * DB * DR * DD * DA. \qquad (3)$$

The infrastructure factor (DI) is radial distance between origin (A) and destination (B) divided by the actual shortest route from A to B over the road network. The inverse of this factor is sometimes used as a correction factor to derive actual distances over the road network from radial distances in models where radial distances are used. We have seen correction factors between 1.1 and 1.7 in empirical work. Some complications may arise here. Since driving time is driving distance divided by driving speed, and driving time is what counts, the optimum route may not be the actually shortest route, but must be seen in conjunction with average speed along the route. We assume, however, that optimum route is shortest route. The infrastructure factor was assessed at 0.83. Potential for improvement is of course zero from the viewpoint of the individual firm.

One haulier computed the infrastructure factor for his situation as follows. With his vehicle planning package, he planned 27 normal trips along the given road network. The total number of kilometers was 7880. When the same trips were planned using radial distances, the total number of kilometers was 5977. Hence the infrastructure factor was 0.76.

The backhaul factor (DB) indicates the loss of efficiency due to the fact that there is not always a return freight. It is defined as 50 per cent + ½ * (the percentage of trips where there is a return freight). In regional breakbulk distribution, the backhaul factor depends on the percentage of trips where, after distribution or in between deliveries, goods are collected to bring back to the depot or transshipment point. If there is a backhaul, the return trip is treated the same as the forward haul. Since collected goods often have characteristics different from distributed goods (e.g. fewer addresses and larger quantities) this may introduce an inaccuracy into the model (Goetschalckx and Jacobs-Blecha, 1989). If this inaccuracy is too severe, the backhaul factor should be integrated with the detour factor, to be discussed below. The backhaul factor in regional distribution of breakbulk is assessed at 0.63, with considerable potential for improvement.

The routing factor (DR) is the famous inefficiency caused by not visiting the given destinations in the optimal sequence. It is the optimal length of the route along the given destinations divided by the actual planned route length. With all the vehicle scheduling software packages available nowadays, it is unlikely that the routing factor can still significantly be improved.

However, the route to be taken along a given number of addresses may be longer than the exact traveling salesman solution for other, more serious reasons than suboptimal heuristics: certain vehicles and even drivers may not be allowed to visit certain addresses; and, most seriously, certain addresses may not be visited at certain times, the problem of time windows. The trouble is, that in practice shippers demand deliveries within more and more narrow time windows and more and more at the same time. Also in urban distribution, the municipal authorities impose more and more narrow time windows.

One haulier computed the cost of time windows as follows. In his situation, 17 per cent of all shipments had time windows imposed on them. Except for the time windows, there was no reason to suppose that these shipments differed from the other shipments. A sample trip planning was made in the usual way, and one in which all time windows were relaxed. The latter saved 18 per cent of the trip costs. In other words, in his situation, time windows doubled the distribution costs for the shipments concerned!

For these reasons, the routing factor is assessed at 0.85 and there is some potential for improvement.

The detour factor (DD) is the straightforward consequence of the fact that shipments are grouped and delivered in roundtrips rather than in full truckloads transported straight between the depot and the first address, between the depot and the second address, etc. Even if the routing factor is 100 per cent, i.e., if the sequence in which the addresses are visited is optimal, the detour factor may be well below 100 per cent. DD is defined as (single distance between the depot and destination i) * (size of shipment i) summed over all shipments i, divided by ((half the total distance of the roundtrip) * (capacity of the vehicle)). Since the return trips are already dealt with by the backhaul factor, here we only deal with the forward trips and we consider half the distance of the roundtrips as forward trips.

The detour factor is inherent in breakbulk distribution, but it is a serious reduction of efficiency. Given the trend of decreasing shipment sizes, the detour factor is getting more serious and there is not much that can be done about it, except for the hauliers to calculate the detour factor into their freight rates and let the shippers decide if they wish to further decrease shipment sizes. Expert opinion assesses the detour factor at 0.62, stable, and moderately liable to improvement.

For an example, consider Figure 1. From inspection, it can be seen that the optimal route is A-B-C-D-E-A of length 19 units. If a vehicle instead takes the actual route A-B-D-C-E-A, the length of the trip is 21 units. The routing factor is therefore 19/21 or 90 per cent. Now consider a truck departing from A and delivering one quarter truckload at B, C, D and E each. If the truck makes this trip, along the optimal route, four times, and covers a length of 4 * 19 = 76 units, it has delivered one FTL at each address. Alternatively, the truck might deliver one FTL by 4 return trips to B, C, D and E each, covering 48 units. Hence the detour factor is 48/76, or 63 per cent.

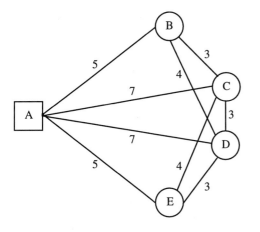

Figure 1. Numerical example of road network, to explain routing factor and detour factor

One haulier computed a somewhat more realistic example. He planned 10 normal trips with 20 stops; the total length was 2696 kms. Assuming that all deliveries were equal to 1/20 of a full truckload, each trip would have to be executed 20 times in order to get one FTL at each address, in total, 53920 kms. Alternatively, the total length of 200 FTL trips to each of the 200 addresses was calculated: 40230 kms. Hence the detour factor for this sample was 0.75.

The actual trip execution efficiency factor (DA) is the ratio between planned trip length and actual trip length. This may be greater than 1, but will usually be below 1, due to unforeseen detours, wrong directions taken, wrong addresses visited, etc. The average trip execution efficiency factor is assessed at 0.88, with quite a large dispersion, and potential for improvement.

For the numerical assessment given here, distance efficiency would be

$$D = 0.83 * 0.63 * 0.85 * 0.62 * 0.88 = 0.24,$$

with some potentialities for improvement, especially in the backhaul (0.63) and trip execution (0.88) efficiency factors.

Partial speed efficiencies

The point of departure for speed efficiency (S) is the maximum speed allowed by law for trucks anywhere in the country.

We distinguish two speed efficiencies, called speed limit factor (SL) and congestion factor (SC):

$$S = SL * SC. \qquad (4)$$

The speed limit factor (SL) is the actual average speed limit over the roundtrip route divided by the maximum speed allowed in the country. For average speed, the harmonic average must be taken. For example, for one third of a roundtrip route a speed limit of 80 kms/h is allowed, for one third a limit of 60, and for one third a limit of 40; the maximum speed for a truck allowed in the country is 80 kms/h. Then the actual average speed limit is 55.4 kms/h and the speed limit factor is 0.69. The speed limit factor is assessed at 0.71, stable, and the haulier can do nothing to improve it.

The congestion factor (SC) is the actual average speed driven over the roundtrip route divided by the average speed limit over the roundtrip route. Assuming that truck drivers generally drive as fast as they can, efficiency losses are wholly due to congestion of the roads, including difficulties to find a parking space. The congestion factor is assessed at 0.75 and is fast decreasing with the road system approaching full capacity utilization. Furthermore, the congestion factor is highly unstable, which makes vehicle scheduling, with narrow time windows and just-in-time deliveries, virtually impossible. No individual haulier can do anything about it. Only concerted action with the government may help.

According to the expert assessment speed efficiency would be

$$S = 0.71 * 0.75 = 0.53,$$

unstable, with no outlook for improvement.

Partial capacity efficiencies

Depending on whether weight, or volume is the binding constraint, capacity efficiency (C) should be broken down in few, or many partial efficiencies. With weight, the series of efficiencies would proceed from maximum gross vehicle weight, via maximum payload weight and various gross packaging weights, to net product weight.

But experts agree, that volume usually is the binding constraint in regional breakbulk distribution. Hence we split up capacity efficiency into seven partial volume efficiencies: capacity factor (CC), floor occupancy (CF), height utilization (CH), pallet load factor (CP), box load factor (CB), net product factor (CN), and actual loading execution efficiency (CA):

$$C = CC * CF * CH * CP * CB * CN * CA. \qquad (5)$$

The capacity factor (CC) is the actual load volume in cubic metres of the vehicle employed as a percentage of the maximum possible load volume in cubic metres permitted by technology (e.g., short coupling) and law (e.g., minimum cabin length, maximum vehicle length and height and width). An average load volume in regional breakbulk distribution may be 40 m3, as opposed to 120 m3 maximum load volume in the Netherlands, giving a capacity factor of 0.30. An increase would hamper the accessibility of the delivery addresses, so there may not be much potential for improvement of this factor. Note that in Sweden the maximum load volume is about 170 m3 for a 24-meter, high-volume truck combination. The expert assessment value is 0.27.

The floor space occupancy factor (CF) defines the percentage of the floor space occupied by freight, say, pallets. If pallet loads are protruding, pallets may not fit nicely onto the truck floor. Floor space may also be lost due to odd-shaped cargo units like machines, carpets, etc. The CF of a fully loaded truck is assessed at 0.83, with some room for improvement.

The height utilization factor (CH) is the average net height of the loading space that is occupied by the freight occupying the floor space. If two pallets of 15 cms thickness are stacked on top of each other with 9 cms manoeuvering space above them, gross height of say, 3 metres, is already reduced by 39 cms, giving a CH of 0.87. If the pallets were replaced by slipsheets, efficiency would improve more than improvement of the famous routing factor can ever hope to achieve. On average, CH is assessed to be very low, viz. 0.47, and variable, and it is deemed worthwhile to investigate possibilities of improvement, for instance, by installing a double floor.

It should be mentioned that hauliers and forwarders are less interested in the efficiencies that follow below. These should be of more interest to shippers, i.e., the ones who pay the freight, is what the hauliers and forwarders say. However, this is an illusion if shippers pay for weight, rather than volume. If shippers pay for weight, they are not interested in volume efficiencies at all. As a result, nobody is interested and many inefficiencies remain unnoticed.

The pallet load factor (CP) is the sum of the volumes of the boxes loaded onto the pallet as a percentage of the imaginary enveloping box around the pallet load, consisting of its projection on the floor space multiplied by its net height (excluding the height of the pallet itself, which is accounted for in the height utilization factor). It is one minus the percentage of 'air' within the enveloping box. It is assessed at 0.69, variable, with some potential of improvement (Ram, 1992).

The box load factor (CB) is similar for the inside of boxes on the pallet, as the pallet load factor is for the contents of the enveloping box of the pallet load. It is the percentage of space in the gross hull of the boxes that is occupied by, let us assume, the final, smallest packaging unit, probably the packaging unit that is sold in the shop. Usually, sales units fit well into their boxes. Hence, CB is assessed at 0.88, stable, with hardly any potential for improvement.

The net product factor (CN) is the percentage that the real, pure product, whose transportation the story is all about, occupies in the smallest, final packaging unit. This percentage may be quite small. For instance, the volume of pure perfume as a percentage of the retail package size may be like 0.10. Commercial people know that selling 'air' can make good profit. The average CN is assessed at 0.39, variable, but for commercial reasons impossible to improve.

The actual loading execution efficiency (CA), finally, is a garbage can efficiency to account for all volume efficiency losses that have not been accounted for elsewhere and that may be caused by the fact that actual stowage is not according to plan. It is assessed at 0.86, unstable, and worthwhile to attend to by operations management.

The capacity efficiency resulting from the numerical assessment reported here would be

$$C = 0.27 * 0.83 * 0.47 * 0.69 * 0.88 * 0.39 * 0.86 = 0.021,$$

unstable, with lots of possibilities for improvement.

Survey of expert assessments

In Table 1, the information gathered from regional breakbulk distribution experts is summarized.

Brushing up all partial efficiencies of the application in regional breakbulk distribution, the result is the overall efficiency (see the model consisting of eqs. (1) to (5)):

$$E = 0.16 * 0.24 * 0.53 * 0.021 = 0.00043.$$

The fact that this is less than one promille need not disturb us. Our theoretical, ideal starting point meant such a long take-off run that not much is left over for the overall efficiency. It is more important to consider possible improvements. If efficiency is increased to one promille, it is more than doubled!

Table 1. Expert consensus, with a cut-off after two rounds, for the case of regional breakbulk distribution by road

partial efficiency[*]	expert assessments		
	average (percentage)	dispersion[**]	potential[***]
TB (business)	37	1.2	3.2
TA (availability)	94	2.8	1.1
TU (utilization)	91	2.4	2.2
TD (driving)	50	3.3	3.5
DI (infrastructure)	83	1.0	1.0
DB (backhaul)	63	2.6	3.4
DR (routing)	85	1.6	2.2
DD (detour)	62	1.4	1.9
DA (actual)	88	2.7	2.6
SL (limit)	71	1.0	1.0
SC (congestion)	75	3.6	1.0
CC (capacity)	27	2.2	1.5
CF (floor space)	83	2.7	2.6
CH (height)	47	4.1	3.1
CP (pallet)	69	2.2	1.8
CB (box)	88	1.3	1.2
CN (net product)	39	2.5	1.0
CA (actual)	86	3.2	2.8

[*]   T = time, D = distance, S = speed, C = capacity
[**]   Coefficient of variation on a five point scale; 1 = very small, 5 = very large
[***] Potential for improvement on a five point scale; 1 = very little, 5 = very much

CONCLUSIONS

We have formulated a general, four-dimensional, physical efficiency model of goods transportation, expressed in the dimensions time, distance, speed and capacity. The overall physical efficiency measure of goods transportation is the continuous product of the four dimensional efficiencies.

We have applied the model to the case of regional breakbulk distribution/collection by truck. We have broken down the four dimensional efficiencies into 18 partial efficiencies. Our starting point was an unattainable, theoretical, ideal situation in which goods are transported continuously, non-stop, along the shortest route, at maximum speed, at maximum vehicle capacity, from A to B and back.

The reason to study this long trajectory is for managers and researchers not to overlook any possible avoidable loss of efficiency and not to focus prematurely and exclusively on certain well-known, partial problems, like the optimal routing problem. The assumption is that sometimes with simple means and measures more efficiency can be gained than with complicated and sophisticated ones.

The usefulness of the breakdown into the partial efficiencies was evaluated and their numerical values were assessed empirically using a semi-Delphi method in two rounds, involving 15 experts from 11 road haulage and forwarding companies in Sweden and the Netherlands. The experts assessed the average efficiencies, their dispersion, and their potential for improvement.

The overall efficiency, calculated as the continuous product of the assessed partial efficiencies, was 0.00043. This is no reason for dispair, because the starting point was a theoretical one and this leaves much room for improvement.

The theoretical results of this study show that a theoretical starting point for an overall physical efficiency model of goods transportation could be found for a full trajectory of partial efficiencies. This prevents researchers and management from focussing on barren parts of the trajectory and overlooking parts where much may be learned or gained.

The empirical results of this study point to certain parts of the trajectory of regional breakbulk distribution/collection that show a low average efficiency and/or a high coefficient of variation, combined with a good potential for improvement (see Table 1). Specifically, we mention: business time (TB), driving time (TD), backhaul factor (DB), actual trip execution (DA), floor space utilization (CF), height utilization (CH), and actual loading execution (CA).

It is the authors' intention to study a number of partial efficiencies in depth. A promising area of further research seems to be stop management - the control of the stopping times of distribution vehicles (the stopping factor being one minus the driving factor). Especially the impact of high-quality information systems (vehicle navigation systems, on-board automatic registration systems, mobile communications systems and the like) and their impact on the various partial efficiencies (driving factor, backhaul factor, routing factor, detour factor, trip execution factor and the like) are a proposed area of further research.

ACKNOWLEDGEMENT

The authors are much obliged to the following experts from road hauliers and contract distribution companies, who contributed to the semi-Delphi exercise, while any errors remain the authors' responsibility:

- Gunnar Alden, Bilspedition Domestic AB, Gothenburg;

- Luud Colsen, Scansped Holland BV, Tilburg;

- Bo Falk, Högsbo Transport and Distribution AB, Gothenburg;

- Ingemar Jönstrand and Bengt Blomqvist, GB Framåt Transport AB, Gothenburg;

- Bo Ireståhl, Bilspedition AB, Gothenburg;

- Jos Koopmans and Piet Lammers, Koninklijke Frans Maas Groep NV, Venlo;
- Hans Malmberg, ASG Sweden AB, Gothenburg;
- Peter van der Meij, Pieter Weekers and Edwin de Brouwer, Van Ommeren Intexo BV, Veghel;
- Roger Nilsson, TGM AB, Gothenburg;
- Sten Weilefors, Göteborgs Lastbilcentral, Gothenburg;
- Sten Åstrand, Bilspedition Halland AB, Halmstad.

ACKNOWLEDGEMENT

This chapter is based on a paper first published in Transport Logistics, vol. 1, no.2, 1997. We are grateful to the editor of this journal for granting permission for it to be reprinted.

REFERENCES

Chow, G., Heaver, T.D., and Henriksson, L.E. (1994), 'Logistics performance: Definition and measurement' *International Journal of Physical Distribution & Logistics Management* 24/1, 17-28.

Clarke, R.L. (1991), 'The measurement of physical distribution productivity: South Carolina, a case in point', *Transportation Journal* 31/1, 14-21.

Foster, T.A. (1989), 'Transportation quality program', *Distribution* 88/8, 74-82.

Gattorna, J. (1988), 'Effective logistics management', *International Journal of Physical Distribution & Materials Management* 18/2-3, 4-92.

Goetschalckx, M., and Jacobs-Blecha, C. (1989), 'The vehicle routing problem with backhauls', *European Journal of Operational Research* 42/1, 39-51.

Heskett, J.L. (1994), 'Controlling customer logistics service', *International Journal of Physical Distribution & Logistics Management* 24/4, 4-10.

Irestȧhl, B. (1984), 'Planning of a production system for domestic, line-based truck transportation' (in Swedish), Technical University of Linköping, Sweden.

Livingstone, G. (1992), 'Measuring customer service in distribution', *International Journal of Physical Distribution & Logistics Management* 22/6, 4-6.

McKinnon, A.C., Stirling, I., and Kirkhope, J. (1993), 'Improving the fuel efficiency of road freight operations', *International Journal of Physical Distribution & Logistics Management* 23/9, 3-11.

Mentzer, J.T., and Konrad, B.P. (1991), 'An efficiency/effectiveness approach to logistics performance analysis', *Journal of Business Logistics* 12/1, 33-62.

Novack, R.A. (1989), 'Quality and control in logistics: A process model', *International Journal of Physical Distribution & Materials Management* 19/11, 1-4.

Ploos van Amstel, M.J. (1990), 'Managing the pipeline effectively', *Journal of Business Logistics* 11/1, 1-25.

Quarmby, D.A. (1989), 'Developments in the retail market and their effect on freight distribution', *Journal of Transport Economics and Policy*, January, 75-87.

Ram, B. (1992), 'The pallet loading problem: A survey', *International Journal of Production Economics* 28, 217-225.

Rhea, M.J., and Shrock, D.L. (1987), 'Measuring the effectiveness of physical distribution customer service programs', *Journal of Business Logistics* 8/1, 31-45.

Ruijgrok, C., Janssen, B., de Leijer, H., (1992), 'LOPER - An instrument to measure the logistical performance of logistical service organisations', *paper presented at the 6th World Conference on Transport Research*, Lyon, France.

Taha, T.T., and Taylor, G.D. (1994), 'An integrated modeling framework for evaluating hub-and-spoke networks in truckload trucking', *Logistics & Transportation Review* 30/2, 141-166.

Tilanus, C.B., (1990), 'What is called degree of utilization?' (in Dutch), *Nieuwsblad Transport*, 18 August 1990.

Walker, I.O., and Wirl, F. (1993), 'Irreversible price-induced efficiency improvements: Theory and empirical application to road transportation', *Energy Journal* 14/4, 183-205.

8

INFORMATION FLOWS ALONG INTEGRATED TRANSPORT CHAINS

Johan Woxenius

ABSTRACT

In transport systems, consignments are grouped to accomplish economies of scale. Parcels are piled upon pallets and the pallets are loaded into containers, swap bodies or semi-trailers that are transported by trucks, trains, ships, or aeroplanes. The information system controlling the goods flow is especially important when general cargo with many small consignments is concerned. In integrated transport chains, unit loads are hauled by rail over the longer distance and are collected and delivered by road. Many organisations are involved, which makes it even more complicated.

The purpose of this paper is to show how information can be grouped in a joint information system and how it is related to the consignments for whose flow it is used to control. Part one of the paper is a theoretical discussion about grouping of goods and information. A short description of integrated transport chains is included. Part two is an application to two real world transportation systems used to show some of the potential savings with a joint information system. The chosen objects of study are one domestic and one border-crossing transport relation. The method used is primarily structured interviews with key personnel of the participating organisations.

Keywords: Combined transport, data communication, information systems, integrated transport chains, intermodal transport

INTRODUCTION

In most transport systems, consignments are grouped to accomplish economies of scale in terminal handling and during line haul. Pieces are put into parcels, parcels are piled upon pallets and pallets are consolidated into bigger unit load units or vehicles. Likewise information is grouped in different ways to control the goods flow - up until today mostly as piles of filled in forms on carbon papers but increasingly as communicating computer systems.

Studies of integrated transport chains (ITC's - see also the enclosed abbreviation list) focus the flow of consignments through links and nodes rather than the involved physical transport resources or modes of transport which often is the case in combined transport or intermodal transport studies. The conception ITC is neither limited to any combination of transport modes nor to the use of more than one mode. However, the basic goal of most research in the field is to transfer goods flows from road to rail and if nothing else is stated, the ITC conception is here used in the meaning of combined road/rail transport of goods consolidated into unit loads such as containers, swap bodies and semi-trailers.

In a typical ITC, the goods are wrapped up and put into a cardboard box that is piled onto a pallet of standardised size. The pallet is then loaded into a likewise standardised unit load. The unit load is finally loaded for transport by vehicles and vessels such as trucks, trains, ships or, occasionally, aeroplanes. Some ITC's are operated by a single company controlling all activities and all necessary resources, but most chains involve many different actors that have to co-operate to achieve effectiveness as well as efficiency. Activities and actors along an ITC are both many and of various kinds which makes the information flow both difficult and crucial to chain performance.

Research on combined transport and ITC's has for long been concentrated upon transshipment technologies and prediction of flows but the administrative system has not yet been thoroughly investigated by researchers. Nevertheless, EU funding within the fourth framework programme widens the ITC research and much effort is currently directed towards administrative processes and information systems (IS's) for ITC's (Cordis, 1996).

Information systems and integrated transport chains are closely related since a good information system is essential for a transport chain to be truly integrated. Consequently, the purpose of this article is to show how groupage of goods and information are, or can be, interrelated. It is not meant to outline futuristic solutions delivering us from all evil, but to discuss some relevant and rather easily realised alternatives to today's business practices. ITC's are chosen as a frame of reference since many consignments and many participating organisations are included. The analysis is confined to information needed for controlling movements of consignments; systems for equipment control are not covered.

Besides this introduction, the article consists of three parts. Part one is a theoretical discussion about groupage of goods and information including a short description of ITC's in order to explain the environment of the IS. In part two, four concepts of information handling in ITC's are presented. Part three, finally, consists of two case studies used for showing some potential savings with better information processing. The chosen objects of study are one Swedish domestic transport relation and one border-crossing transport relation.

The theoretical part of the article is basically of a conceptual nature based upon own thinking and a wide variety of books, articles and brochures. The primary data for the case studies were gathered through structured interviews with key personnel of the participating organisations. The view of favouring conceptual thinking before detailed analytical research on information systems is supported by Checkland (1988):

"The whole field of information systems (...) lacks its Newton to bring it conceptual clarity."

This work is part of a combined transport research project that commenced in 1991. The research is financially supported by Swedish State Railways and includes topics such as industrial organisation, production systems, traffic designs, commercial and technological openness as well as technology surveys. The present article is an extended and partly up-dated continuation of an earlier paper (Woxenius, 1994/a).

GROUPAGE OF GOODS AND INFORMATION

Groupage of goods and information faces different realities according to Bollo *et al.* (1992):

"Transport and the telecommunications systems possess similar organisation with nodes and links that lead to numerous interface problems. The main difference lies in the fact that breakdown and consolidation are now automatically handled for communications, not for goods!".

This is partly true, but sorting and consolidation of mail and parcels are today heavily automated. Technology advancements such as the French automatic rail/rail transshipment terminal Commutor and the Delta/Sea-Land sea/road container terminal in Rotterdam also point to a possible future of automatic handling of larger goods units. The truth is rather that automatic data processing is a prerequisite for automatic goods handling.

Groupage of goods

Groupage enables consolidation of goods, from the tiniest piece to the largest unit, into efficient transport flows where economies of scale are present, both in handling and during line haul. The whole system must be standardised to some extent so that all units fit into larger units in the hierarchy in efficient patterns. A striking example is the size of Swedish sugar cubes that was changed some years ago to enable more efficient piling into packages that in a pattern can form standardised 400 * 600 mm modules. Shelves in shops are designed to fit these modules, and patterns of modules and sub-modules efficiently fill the 800 * 1 200 mm measures of EUR-pallets. The EUR-pallets are handled with fork lift trucks and fit into lorries and unit loads, the ISO-container excluded. An example of a groupage chain is shown in Figure 1.

Figure 1. A groupage chain

The size of consignments is crucial to information processing and particularly to the tracking and tracing activities, i.e. activities to keep control over the goods flow and find consignments deviating from the transport plan. A consignment is here defined as the smallest loose unit that is handled at transshipment nodes. Consequently, the consignment size ranges from a single spare part in an express parcel service to 500 000 tons of crude oil shipped overseas. The size of consignments also varies along a transport chain due to consolidation and breakdown activities.

The core of the ITC's treated here spans over two combined transport terminals with one or more rail links between them. Following the definition above, unit loads such as containers, swap bodies and semi-trailers, constitute the consignments. Smaller consignments inside the unit loads are in this paper referred to as subconsignments, while transport units with two or more unit loads, e.g. trains and ships, are called superconsignments.

A traditional ITC contains a pick-up service with a short road haulage from the place of dispatch to a combined transport terminal, transshipment to a rail wagon, long distance rail haulage, another transshipment and, finally, a delivery road haulage to the consignee. The whole idea of combined transport implies that many different types of activities, resources and actors are involved in transporting unit loads from the consignor to the consignee. This network approach is shown in Figure 2. The arrows indicate normal supplier relations between the actors which naturally also affect the information system.

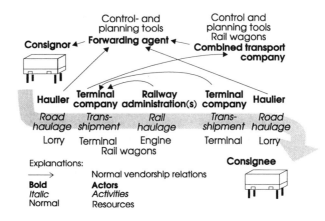

Figure 2. An integrated transport chain with normal supplier relations

The actors involved are typically a forwarder maintaining contacts with the end customer, an SME (small or medium sized enterprise) haulier, a terminal company, a combined transport company and one or more railway companies. A deeper system analysis including the industry structure is presented in an earlier paper written within the project (Woxenius, 1994/b).

Groupage of information

The value of transport information was clear to John D. Rockefeller in the 1870's when he conquered the US oil industry. Contracts with the Pennsylvania, the New York Central and the Erie railroads stipulated that he and his partners had right not only to substantial freight rebates for own shipments, but also to the rebates of competing oil companies. Even more important, they where entitled to receive copies of their competitors' freight lists including information about their customers, products, prices and terms of payment. By use of that information he threatened to consequently underbid them and thus forced them to either co-operate with Standard Oil or run out of business. Within less than three months, 21 out of 26 competing refineries ceased their operations in Cleveland (Dillard, 1967).

The value of information has further increased during the last 125 years due to the increasing complexity of society in general, particularly in connection with trade and logistics. Advanced and computerised information systems become increasingly important in line with new spatial trade patterns as well as demand for high quality transportation of small consignments. Large parcel services like Federal Express, DHL and UPS need extensive data processing to operate their networks. The fact that Federal Express is among the 10 biggest customers of IBM (Fredriksson and Holmlöv, 1990) indicates this importance. Besides parcel services, single-mode operators such as railways and road hauliers have for long used in-house IS's supporting their operations.

Many different, however similar, documents are today used for managing goods flows in ITC's, a fact resulting in substantial parallel information flows. Unfortunately, the computerised IS's of the different transport organisations hardly speak to each other - information is still transferred between actors by use of telephone, telex and telefax and even mail (see e.g., Kanflo and Lumsden, 1994 and Andersson et al., 1993).

A lot of time and effort is spent today on the repeated feeding of pieces of information from consignment notes (CN's) and customs declarations into the actors' different computer systems. Furthermore, the risk of expensive errors is high. However, implementation of EDI systems is bridging the gap, however only at a gentle pace. For instance, Swedish EDI installations often involve the supplier and the customer, but seldom the transporting company which still uses printed documentation (Kanflo and Sjöstedt, 1992). This is especially true for transport between production units within one company.

The slow pace of implementation does not mean that the problem is not being addressed - the transportation industry has traditionally been very active in developing standardised messages (see, e.g., Fabre and Klose, 1990). Out of the totally 125 UN/EDIFACT messages drafted or standardised by December 1992, 25 were dedicated container messages and some additional 30 directly concerned the container industry (Schlieper, 1992).

The shipping industry is a good exception from the slow implementation of EDI in the transportation industry as a whole. Especially Swedish shipping lines have a long tradition of innovation in information handling, with the basic attitude that the traditional bill of lading was inefficient. In 1966, the England-Sweden Line introduced its ESL-document which was a new type of bill of lading that was printed when the load unit entered the ship. Atlantic Container Line (ACL) used its Datafreight Receipt issued with the bill of lading but wired over the Atlantic to reach the destination port prior to the cargo. Thus arrangements for hinterland transport could be made in advance. Broströms introduced the document that finally replaced the bill of lading as a legal document. Its Liner Waybill was actually a CN for liner shipping.

Spera (1987) states three tasks that IS's in the logistics field have to fulfil:

1. The ability of making quick arrangements

2. The guarantee of data protection

3. The fulfilment of legal conditions

Quick arrangements along an ITC is taken care of by use of booking systems for various resources and activities, usually operated by the forwarders or railway companies. The booking system has to initiate action at all actors to reserve capacity for the consignment along the entire ITC.

The possibility to *guarantee protection of data* is essential. In the USA, business information in wrong hands has caused transport companies severe losses (Gellman, 1994). The challenge is not only to distribute information steadily to many actors, the question of restricted data access must be dealt with seriously. The "road based" forwarders' IS's contain information of the transport customer that they fear will be misused by their long time competitors, the railway companies. Transport information has long been available through CN's, but the problem is magnified with powerful analysis tools used in combination with the accessibility of computerised IS's.

The fulfilment of legal conditions is basically a question of adapting law to new information technology. For instance, Swedish law no longer requests a CN for domestic transport, but if it is issued, it has to meet certain standards. Border crossing transport is far more complicated, due to customs requirements and, unfortunately, also due to some outmoded traditions within European transport industry. In Germany, for instance, the law requires that all CN's are physically kept by the engine driver. Further harmonisation within the European Union will heavily affect the development of legal requirements of transport information, however requiring substantial time and effort by all involved parties.

Ideally, information should be keyed, or automatically read, into a computer system once and then automatically processed to enable resource reservation throughout the transport chain, transport arrangement planning, tracking and tracing, announcements of deviations from the plan, advising the customer and finally invoicing the performed transport service. Every change of status should be recorded, preferably by auto-ID antennas along the path. Statistics necessary for system improvement should also be transferred and processed automatically. Every actor should electronically receive the information needed, and only the information needed.

The great advantages of computerised IS's is not fully utilised if only the information of today's CN's is transferred electronically. The great potential lies in letting the information flow run parallel with the goods flow and enable it to be used to plan and control activities (see, e.g., Polewa and Lumsden, 1996). Nevertheless, this article does not aim to present futuristic solutions but realistic alternatives to today's practice. Moreover, many attempts of business wide solutions such as electronic freight exchanges have failed to reach a wide market use (Henningsson and Isacson, 1994).

Computerised IS's are costly, especially when the aim is to link different in-house systems. The actual potential of implementing a computerised IS must thus be thoroughly investigated (see, e.g., Cabello, 1996). The technocrat's view of solving problems is not always the best - the real need for advanced solutions in combined transport can rightfully be questioned. A European train typically has a capacity of less than 50 unit loads and a double-stack train in the USA less than 300 containers. These are no large numbers and, therefore, advanced IS's are not critical to the core ITC operation. However, the rail link is only one part of a wider transport system which needs an uninterrupted chain of extensive information support to work properly. This view is shared with Gronland (1987):

"... the contribution from a subsystem should be controlled according to the criteria set by a higher level system. (...) If this is not possible our control problem is not how to control a given transport, but how to change from one transport to another."

The core of the ITC can clearly be treated as a subsystem in the eyes of the forwarder responsible for door-to-door transport. The growing interest in the ITC conception focusing the goods flow rather than the involved transport modes also points towards benefits from connecting the IS's.

FOUR INFORMATION FLOW DESIGNS FOR ITC'S

In this section, four concepts of information flow design for controlling ITC's are identified, of which the first two use paper as a medium and the latter two computer communication. Consignments and subconsignments must in all concepts be marked with an ID tag to enable manual control and tracking.

Hierarchical consignment notes

For competition reasons, the road transport based forwarders with domestic general cargo do not have to, or want to, hand over information of all subconsignments to the railway companies. Instead, they issue another consignment note (CN) that only specifies the physical unit load while the CN's covering the subconsignments are kept inside the locked unit load all along the ITC. Hence, the original CN's physically accompanies the subconsignments, while other CN's (or equivalent documents) are issued for unit loads, wagons, wagon groups and so on. In other words, the documents concerning the individual consignments are not available on the next hierarchical level. The concept is graphically presented in Figure 3.

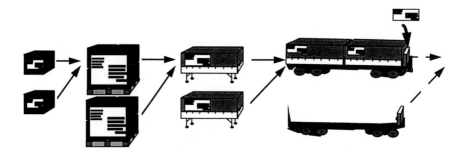

Figure 3. Concept 1: Hierarchical consignment notes

Only prospective customs declarations and information about hazardous goods have to be handed over to the next actor; other information about the subconsignments is kept inside the transshipped load unit. A tracking operation then has to be made either manually, which is not realistic with consignments stuffed tightly together in unit loads, or hierarchically supported by use of paper or computer based IS's of the actors. The tracking operation must then include all actors from the start of the chain to where the subconsignment is physically located.

The procedure described above is the default system for rail based ITC's today. The CN's are not suitable for controlling a complex transport system and, hence, the different actors often use parallel, in-house, IS's to control their individual operations. CN's can serve as a medium for information transfer, but information is also gathered over the telephone or telefax. The key disadvantage is that repeated data entries are very time-consuming and lead to many errors; also information is not transferred prior to the transport.

Groupage to the next consignment level

As mentioned, customs clearance and transport of hazardous goods demand documents to be available for inspection close to the consignments along the whole ITC. The original CN's covering the subconsignments are then collected and transferred to the next actor in the transport chain. ID-tags or copies of the CN's are attached to the subconsignments while the original CN's are put into a document box on the unit load. At the combined transport terminal, these documents are collected and put in the document box of the wagon and later handed over to the engine driver. Documents are thus grouped in the same way as the consignments they are controlling. Copies of CN's usually remain at every hierarchical level, but the CN's are primarily linked to the subconsignments via ID tags. The groupage operations are shown in Figure 4.

This procedure implies two parallel groupage operations, one for goods units and one for documents. Tracking of subconsignments is possible at the highest hierarchical level where all information is concentrated. After locating the proper CN, the subconsignment can be referred to a specific wagon group, a wagon, a unit load and so on.

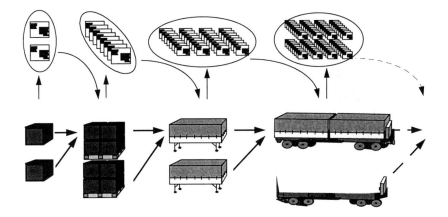

Figure 4. Concept 2: Groupage to the next consignment level

Border-crossing ITC's are traditionally controlled in this way with repeated collecting of documents. However, to offer faster transport services, international combined transport operators demand customs clearance at the destination terminal or at the consignee, enabling the trains to pass all borders without delay.

Communicating in-house computer systems

The third alternative information concept presented here is similar to the first concept but no printed information accompanies the consignments. The forwarder or haulier assigns a subconsignment to a unit load, the terminal operator assigns a unit load to a wagon or a group of wagons and the railway company connects the groups of wagons to a train. Every actor retains information about the contents of units transferred to the next hierarchical level. Data is entered once, manually or automatically, and is then communicated between IS's similar to electronically transferred CN's. Communication is preferably made using EDI since this open interface environment involves the IS's of many different actors. Figure 5 presents the communication principles.

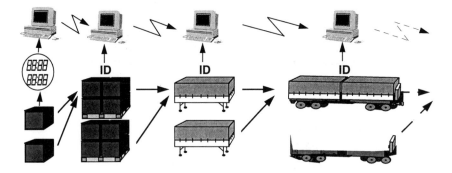

Figure 5. Concept 3: Communicating in-house computer systems

Tracking is made hierarchical by position questions between different computer systems. This can be accomplished by the implementation of EDI-links between all companies. Thereby, routines for search clearance are of a dual nature and thus easier to control. Documents that according to law have to be printed are printed from the computer systems directly. The concept is similar to that presented by Jarke (1982) who lays the foundations for a hierarchical distributed decision support system (DSS) for a sea/rail/road container transportation system.

Many initiatives are now taken to improve and spread the use of computer communication to a wider range of transport companies. Some years ago, the European Commission financed a research and development project called CombiCom, run by a five-country consortium lead by Cap Gemini of Italy. Unit loads were equipped with standardised ID-tags and combined transport terminals with EDI connections to consignors, hauliers, forwarders and consignees. Data was gathered at the terminals and at certain auto-ID checkpoints along a dedicated test corridor between the Netherlands and Italy (CombiCom, 1994). Subconsignment tracking was not supplied. Unfortunately, the project is no longer supported by the EU, reportedly due to disagreements on how to commercialise the concept. The auto-ID system was supposed to be publicly available, but in practice the solution was limited to work with Cap Gemini software controlling the whole ITC.

In the USA, combined transport involves fewer actors giving fewer interfaces. Contrary to the current situation in Europe, most railroads offer EDI services. Union Pacific has its UPINFO, a PC program for communication with their TCS mainframe computer system enabling wagon ordering, scheduling, tracking and automatic billing. Norfolk Southern has Thoroughbread Quickfo and Burlington Northern has BN Lynx with similar features.

A central data base system

To rationalise information processing, the actors can agree upon using a dedicated IS. The central data base can be administrated either by an independent information broker, such as described by Dubois and Gadde (1989), by a main actor or jointly by all involved actors. Subconsignments are used as information units all the way through the transport chain, and links to transport equipment are accessible. Since a dedicated system is used, communication does not have to be made using EDI. One big advantage is that the exact information needed can be transferred without having to produce several different documents containing similar information. The data base concept is shown in Figure 6.

Realisation of this concept depends to a great extent upon the actors' willingness to co-operate. The potential savings must be emphasised and the savings must be fairly distributed between them. Furthermore, developing the IS is a challenge that, besides program development skills, demands thorough knowledge of business practices. A system approach is thus needed to accomplish the best performance.

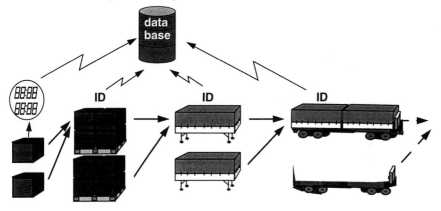

Figure 6. Concept 4: Central data base system

General Electric's Cargo*Link/CTS is one of several common consignment data bases supplied independently of transport operators. It is open to all transport modes and covers tracking of subconsignments from consignor to consignee and is thereby suitable for ITC's. These "do-it-all" solutions have not found the broad application first anticipated. This is partly due to the fact that forwarders see information handling as their core business and they see general systems as threats to their market positions. Also shippers, the end customer of transport services, are typically suspicious to new solutions since they fear to be dependent upon one supplier of transport services.

With a long history of problems, the German State Railways (DB AG) now uses DISK, a new IS for combined transport. Besides being used for operational arrangements at terminals, DISK supports strategic planning and production management. Subconsignments, unit loads, road vehicles and railway wagons are tracked by the mainframe computer in Nürnberg, saving time and effort compared to the other concepts. The system contains sub-systems for reservation of transport of subconsignments and unit loads. DB's own communication network is used for connecting the major ITC actors, including combined transport companies, independent terminal operators and wagon leasing companies. Invoicing is not done by DISK itself but information is electronically transferred to the administrative system. About 65 German combined transport terminals were connected to DISK in 1994. Even international companies are connected and the vision is to extend the system enabling automatic tracking and tracing all over Europe (German State Railways, 1992).

Large schemes like the German DISK have been drawn up in order to integrate the IS's of the national railways, but now when they face competition not only from new small railway undertakings but also from each other, the integration process is slowed down significantly if not halted completely (Sandberg, 1994). The reasoning is simple: even though no more information than today will be accessible, the data bases make the information easy to search and analyse in order to find the competitors' large customers.

CASE STUDIES

The first case is a domestic ITC with general cargo shipped in semi-trailers illustrating a possible change from concept 1 to concept 3. The second case is an international ITC where pulp is shipped in swap bodies. A central data base is proposed to replace an extensive and redundant document flow, i.e., a proposed change from concept 2 to concept 4. Field studies were performed together with a group of graduate students.

ASG Linje 1, domestic freight transport

ASG Linje Ett is a group of hauliers for Sweden's second largest forwarder, ASG, on the domestic Stockholm-Gothenburg connection. Annual flows total 160 000 consignments, equalling 120 000 tons, 90% of which refer to combined transport. Semi-trailer is the dominant unit load although some over-sized containers are used. An additional, articulated road vehicle of 24 meters is used for backup and express deliveries.

The Swedish domestic CN contains 36 information fields to be filled in on carbon paper, giving three additional copies. Neither the combined transport company Rail Combi (RC) nor Swedish State Railways (SJ) uses the CN's - they are physically transported by the backup lorry to the office of ASG Linje Ett in Stockholm.

Figure 7 shows two separate computer networks that communicate electronically internally, but externally via the telephone, the telefax or the mail service. Typically, the road based forwarding agent and the haulier have one common IS and the combined transport terminal company and the railway company another. RC was until 1st of July 1992 a business unit within SJ and RC still uses the mainframe computer of SJ for tracking and arranging train haulage. Nevertheless, most planning activities are made without computer support. A new IS is now implemented.

The interviews conducted show that the CN's of a subconsignment contain much unused information as shown in Table 1.

The consignor is supposed to fill in many pieces of information not used by any other actor since other IS's are used to handle the same information. The information actually used by transport companies, 5 to 6 fields in all, is acquired by telephone or telefax when the consignor books the transport. A pick up order is transmitted within the computer network of ASG and wagon space for combined transport is booked depending on the total number of semi-trailers needed. Telefax is used as medium.

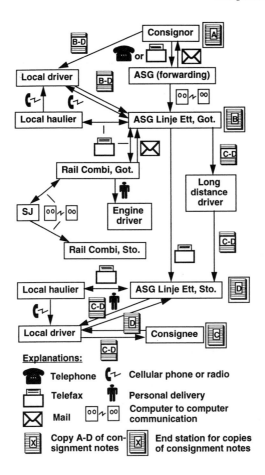

Figure 7. Information media and flow of consignment notes in a domestic ITC

Table 1. Number of information fields of a domestic CN used by actors (out of 36)

Consignor	28
Pick-up driver in Gothenburg	4
ASG Linje Ett (hauliers)	4
Rail Combi (CT operator)	0
ASG (forwarder)	7
Delivery driver in Stockholm	3
Consignee	<u>4</u>
Fields used by no or only one actor	25

Many fields of the CN handle legal arrangements, such as payer and terms of payment. This is normally regulated in long-time transport contracts, making these fields redundant since the consignor states contract number. It is remarkable that the drivers only use information concerning the consignment; address information is supplied by ASG's IS. Information of unit loads and which wagons they are loaded upon is keyed into the SJ data base. The information is used for making an electronic CN stored in the data base. Like the CN of the wagon group, a final train specification with information of train length, weight, hazardous goods etc. is printed and handed to the engine driver.

Problems of connecting road and rail based companies, similar to the interface problems discussed by Spera (1987) and Jarke (1982), are clearly shown in this case study. CN's are filled in by consignors, but the information is not operationally used by transport companies. This information flow shows most similarities with concept 1, although CN's are electronically transferred in the core ITC. Moreover, the subconsignment CN's are not put into the semi-trailers, but still, the road based companies keep the CN's beyond the railway sector's reach. The CN is outdated or badly used and is foreseen to be replaced by an ID tag or a bar code linking up with information in data bases. Since ASG and SJ have IS's developed for single-mode transport, concept 3 with communicating computers is considered best. At a low cost, information processing is then considerably improved without interfering with the single-mode operation.

Skandi, international freight transport

Skandi is a dedicated swap body forwarder in the Danish AP Möller group, selling transport of full swap body consignments. This case describes document handling controlling the transportation of 11 000 annual tons of pulp between Värö and Milan via the combined transport terminals in Gothenburg and Luino. The customer, Södra Skogsägarna (Södra) is a substantial transport buyer. In 1991, 6.2 million m3 of timber were transported to their mills at a cost of 40 million USD. Outbound, 870 000 metric tons of pulp and 334 000 m3 of wooden products were shipped at a cost of 25 million USD.

Södra buys door-to-door transport from Skandi, who in turn buys rail haulage from Intercontainer-Interfrigo, ICF, a combined transport company owned jointly by European railway companies. The wagons, owned by ICF, are consolidated in Copenhagen into a Skandi full-train connection to Luino. Suppliers of ICF are Rail Combi (RC) for terminal handling in Gothenburg, the railways SJ, DSB, DB, CFF and FS for rail haulage, and DanLink, jointly owned by SJ and DSB, for ferry crossings. The terminal in Luino is operated by Skandi's subsidiary Visnova. Road haulage is supplied by closely connected hauliers in Sweden and Italy, but 20% of the consignments are transported by the haulage department of Visnova. This description is an indication of the complex situation in border-crossing ITC's.

Consequently, many transport documents are issued, some of which in eight copies. Printed driving commissions are used by the hauliers; an international CN, the CMR, is used by Skandi for door-to-door transport; and an additional CN, a hand-over note, is used by ICF for rail transport. For customs use, an export document, a certificate of origin and a trade invoice are issued. With full swap bodies, these seven documents refer to the same consignment. The real goods invoice is electronically transferred to Södra's trade office in Milan. Neither Skandi nor Södra communicates electronically externally, although Skandi sends diskettes to RC and

ICF. Furthermore, the railways and the ferry line use wagons, wagon groups and the Skandi full-train as superconsignments. CN's are issued for all these hierarchical levels.

This extensive document flow causes serious problems. First, the documents have to be filled in. Software for laser printing of whole export documents originating from in-house data bases are now being marketed by, among others, IBM Svenska AB, whose ELOGE program prints documents directly, or sends them remotely by telefax or EDI. Skandi's IS prints data on document forms whereby this problem is reduced. Second, and most importantly, keying the information into seven different data bases involves extensive work with a considerable error probability factor. Any mismatching of information creates problems when it comes to customs clearance.

The five main documents have many information fields in common, causing a great deal of redundancy work and risk of expensive errors. The necessity of redundancy is argued by Jarke (1982):

"... each message must include redundant information, in case a message is lost or a special event such as damage or theft of cargo is not communicated separately".

Nevertheless, the need for several documents with similar content could hardly be justified. Consequently, the savings potential for a computerised IS is substantial. A future joint data base contains all information needed and each actor, including the customs, can search for the pieces of information needed. Table 2 shows the redundancy information when all documents are properly filled in.

Table 2. The number of equal information fields in five transport and export documents

		Redundancy
Present in: all five documents	6	6 x 4
four documents	4	4 x 3
three documents	4	4 x 2
two documents	12	+ 12 x 1
Total number of redundant fields		56
Total number of information fields:		174

Customs constitutes a main obstacle when it comes to making the information flow more efficient. SJ hardly uses any CN's in domestic rail transport, but in accordance with international law, CN's are printed from information gathered in SJ's mainframe computer at frontier stations. Swedish customs offers EDI connection to their customs clearance system TDS. However, in this case, a customs commission agent sees to it that the printed documents are stamped at the Swedish border. The European single market gives an option to declare the pulp in Denmark. Visnova, however, takes care of this in Italy, even though Italian customs seem reluctant to adopt more efficient information handling. EDI systems like TDS operated by the Swedish customs, an investment of USD 40 million, are considered to take time to implement in this transport corridor. Tracing services are supplied by DanLink Info in Copenhagen, jointly owned by DB, DSB and SJ. Skandi connects their swap bodies to wagons that are traceable in the IS of DanLink Info. Also ICF has a large development scheme called Euronet with the purpose to feed their central database with information from decentralised data capture points for later printing of documents (Euromodal, 1995).

It is shown that documents used in international transport contain many similar information fields. Due to many actors with different languages, business cultures and computer maturity, clearly understandable guidance information must follow the consignments for a considerable time. However, customs information needs will diminish with the European single market, and the different documents needed can be produced from data in a central data base. It is shown that there is a large savings potential in document handling.

Concept 4 with a central data base should be a lucrative solution to the extensive document flow with repeated manual data entry. In contrast to the domestic case, the forwarder is not road based but a dedicated combined transport user. The data base should then be designed for multi-mode and multi-actor application. The DanLink Info system could be extended to include information on consignments and subconsignments, not only superconsignments. With commercial data included, this should satisfy customs and shippers, giving a complete IS that can replace most documents. Implementation is merely a business decision between companies and authorities.

As long as all these documents are legally requested, the best thing to do is to print them directly from the same database. A Swedish program that supports automatic laser printing of documents is supplied by IBM. The ELOGE export system can print documents directly, over the telefax or through EDI connections.

Data access management presents no real technical obstacle, but many actors state that fear of electronic burglary is a hinder to such an application. However, all actors are currently waiting passively for others to make the expensive mistakes (Kanflo and Lumsden, 1994). There is no doubt about the extensive savings potential involved for actors who replace physical paper flow with light pulses in fibre optic wire.

CONCLUSIONS

Integrated transport chains are very complex due to the large number of operators involved in both the physical transport chain and the market system. The implementation of information systems has primarily focused on single-actor or single-mode transport. However, the time has come for combined transport to enter the computer age, leaving the abundant quantity of printed documents behind. Due to the fact that groupage into unit loads limits the number of consignments, ITC's based upon unit loads do not necessarily need advanced IS's like parcel services. Nevertheless, the railway core of the ITC is only one part of a larger transport system and must be integrated in the total IS. This research points at substantial savings when implementing computer communication, either using EDI between in-house systems or a joint, dedicated data base system.

The difficulties in implementing new information systems linking different companies should, however, not be underestimated. There seems to be a tendency of waiting for other companies to make the expensive mistakes, but at the same time a fear of being left too far behind. The key to success is thought to be to solve the question of distributing efficiency gains between the SME hauliers and the larger forwarders and railways. Today, the former are normally supposed to invest in distributed information systems, while the latter benefit most from it.

ABBREVIATION LIST

Very well known abbreviations as well as abbreviations only used once are not put on the list.

ASG	The abbreviation is used as company name	IS	Information System
CN	Consignment Note	ITC	Integrated Transport Chain
DB	Deutsche Bahn AG (German State Railw.)	RC	Rail Combi AB
EDI	Electronic Data Interchange	SJ	Statens Järnvägar (Swedish State Railw.)
ICF	Intercontainer-Interfrigo	SME	Small or Medium sized Enterprise
ID	IDentification		

REFERENCES

Published references

Andersson, U., Björkbäck, C., Björkman, M., Carlsson, C. (1993). *Informationsflödet i kombinerade transportsystem (Flow of information in a system for combined transport)*, Depertment of Transportation and Logistics, Chalmers University of Technology, Gothenburg. In Swedish.

Bollo, D., Hanappe, P., Stumm, M. (1992). *Standardisation in Information Systems for Commodity Transportation*, Proceeding of the 6th WCTR, June 29 - July 3, Lyon.

Cabello, C. (1996). *Returns on IT investment: the end of the productivity paradox?*, The IPTS Report No. 3, April, pp. 10-13, European Commission, Joint Research Centre, Brussels.

Checkland, P. B. (1988). Information Systems and Systems Thinking: Time to Unite?, *International Journal of Information Management*, 8, pp. 239-248.

CombiCom (1994). *Final Report CombiCom 1*, report about a research project within the European Union research programme Advanced Telematics in Transport, Brussels.

Cordis (1996): see *Unpublished references* below

Dillard, D. (1967). *Economic Development of the North Atlantic Community Historical Introduction to Modern Economics*, Prentice-Hall, Englewood Cliffs, NJ.

Dubois, A, Gadde, L-E (1989). *Information Technology and Distribution Strategy*, IMIT, Working paper No. 1989:3, Presented at the Second Workshop on Information Technology and Business Strategy, Brussels, June 12-13.

Euromodal (1995). *Objective electronic data transfer: ICF-Euronet*, No. 2/1995.

Fabre, F., Klose, A. (Eds.) (1990). *Cost 306 - Automatic transmission of data relating to transport,* Final report, Office for Official Publications of the European Communities, Luxembourg.

Fredriksson, O., Holmlöv, P. G. (1990). *Framgångsrik användning av informationsteknologi inom distribution av varor & tjänster (successful implementation of IT within distribution of goods and services),* TELDOK-report 57, Stockholm. In Swedish.

Gellman, A. J. (1994). *North American Intermodalism,* Swedish Transport Research Conference, January 12-13, Linköping.

German State Railways (1992). *DISK: leistungsfähiges Datensystem für den KLV von morgen (capable computer system for the combined transport of the future),* information pamphlet, Frankfurt am Main.

Gronland, E. (1987). *Some Aspects of Practice: Control in Logistics,* In: Control in Transportation systems, IFAC Proceedings series, 6, pp. 105-106, Pergamon Press, Oxford.

Henningsson, N., Isacson, S. (1994). *Transportbörs i trailer- och containerbranschen (Proposal of a haulage exchange for semitrailers and containers),* Master's thesis No. 94:13, Department of Transportation and Logistics, Chalmers University of Technology, Gothenburg. In Swedish.

Schlieper, H. (1992). *Introduction to UN/EDIFACT - Messages and Framework,* IBM Germany Information Systems, GmbH.

Jarke, M. (1982). Development Decision Support Systems: A Container Management System, *International Journal of Policy Analysis and Information Systems,* 6, pp. 351-372.

Kanflo, T., Lumsden, K. (1994). *Informationsflödet i transportkedjan (The information flow in the transport chain),* Chalmers University of Technology, Gothenburg. In Swedish.

Kanflo, T., Sjöstedt, L. (1992). *IT/EDI in the Swedish Transport Sector, The Workshop on IT - Computer Networking in and Between Economic Sectors - Distribution and Transport,* 5-6 February, OECD.

Spera, K. (1987). *Some Aspects of Practice: Data Processing in Logistics,* In: Control in Transportation systems, IFAC Proceedings series, 6, pp. 107-108, Pergamon Press, Oxford.

Woxenius, J. (1994/a). *Groupage of Goods and Information in Systems for Combined Transport,* Presented at the 7th IFAC Symposium on Transportation Systems, Tianjin, China, August 24-26. In: Liu Bao, Blosseville, J. M., (1994) *Transportation Systems: Theory and Application of Advanced Technology,* Elsevier Science Ltd, Oxford, UK.

Woxenius, J. (1994/b). *Modelling European Combined Transport as an Industrial System,* Licentate Thesis, Report 24, Department of Transportation and Logistics, Chalmers University of Technology, Gothenburg.

Interviews and unpublished references

Börjesson, Willy, VWB Åkeri AB, Gothenburg, December 12th, 1995.

Carlsson Reidar, president ASG Linje Ett, August 27th, 1994.

Cordis (1996). Community R&D Information Service, on-line information service, world-wide-web address: http://www2.cordis.lu/cordis/cord5000.html.

Edvardsson, Göran, Södra Shipping, Växjö, July 5th, 1992.

Polewa, R., Lumsden, K. R. (1996). *The added value of information*, unpublished working paper, Department of Transportation and Logistics, Chalmers University of Technology, Gothenburg.

Randvik, Stig, Randviks Transport AB, (former partner in ASG Linje Ett), Gothenburg, December 12th, 1995.

Ronnebro, A.; Skandi Sverige AB, interview in Gothenburg, March 29, 1993.

Sandberg, B. (1994). *Utvecklingen inom informationshanteringsområdet (Developments in the field of information handling),* Speach at the Swedish Transport Research Conference, January 12-13, Linköping.

9

INFORMATION AS A VALUE ADDER FOR THE TRANSPORT USER

Roudolphe Polewa
Kenth Lumsden
Lars Sjöstedt

ABSTRACT

Many of the developments in logistics in recent years are connected to the need for more information on the goods flows. Numerous tools have been developed that target this particular need and purely aim at increasing cost efficiency in existing control and planning systems. Few if any of these systems have the ability of assessing the value that more information can add to the transport task, seen from the shipper's viewpoint. The aim of this conceptual study is to assess the added value of information in transport operations for the user of the transport system by using the principles of tracking and sequencing of individual transport units.

The theoretical background of the study stems from the insight that information can in many ways be likened to a product. The value of a product increases on the production chain, from the raw to the processed material phase. By the same token the value of a piece of information increases as the raw data, unprocessed information, is transformed into knowledge through meaning- or value-adding. This knowledge can be considered as the final step of the value-adding process of information and it is ready to be used, just as the final product is considered as the entity ready to be handed over to the customer.

Keywords: Contract time, link time, node time, buffer time, consolidation time gain, deconsolidation time gain

INTRODUCTION

Many information systems have been developed to help in tracking transport resources such as vehicles and individual transport units (see for example Bonsal, 1987). These systems usually consist of a transponder or equivalent equipment attached to the vehicle or the transport unit. This equipment can send data in the form of electronic signals to a receiver. These signals can in turn be read at known locations at a determined time or with a determined frequency (Lumsden, 1990).

The problem is that the information thus gathered at various locations might either be redundant at some point of the transport chain or might come too late to be used to the fullest by its recipient, be it the shipper or the user of the transport service. This leads us to define the scope of this study as follows: Information has to be evaluated in relation to and as early as possible on the transport chain in order to swiftly draw advantage of its meaning. There are in view of this two issues to address. The first is how we define the meaning of information, i.e. how we give value to information. The second is which gains can be drawn from the early gathering of the information.

In this study it will be of great importance to define how Information Technology can improve the time utility for the shipper not only by creating automated mirrors of already existing routines but also by identifying or changing the physical processes with the ultimate goal of improving these processes. The first part of the study is devoted to some general aspects pertaining to information valuation. The second part deals with information valuation from the perspective of the logistics chain and touches such aspects as information gathering and information gains. The third part looks into how time gains can be translated into utilities for the customer of a transport service. The last part concludes the study with some final and general remarks.

FRAMEWORK FOR INFORMATION VALUATION

The information value chain

One can identify three concepts that are used interchangeably to characterize the same thing namely data, information and knowledge. A closer look at these concepts shows that they appear successively on a chain that can be called the information value chain. In its initial form information is data. This data can be stored in libraries, in databases or any other medium that does not give them a structure per se. Data in this form has almost no meaning. If the data is structured it becomes information in the sense that it can be communicated. In its communicable form information can also be analysed, interpreted or modelled. The result of this transformation is knowledge. Knowledge is thus information that has been given meaning by its user through analysis, interpretation or modelling. The chain just described can be summarised as a progressive process (Figure 1), starting from the initial data with limited meaning to a more sophisticated piece of knowledge that can be used in a later process.

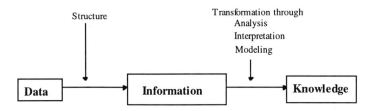

Figure 1. The information value chain

Attributes of information

Information or knowledge can be compared to a product that evolves from its raw form (raw material) to more sophisticated and valuable forms through various processes. By the same token information can be said to be a product in the sense that its transformation from raw data to knowledge for its users augments it gradually with meaning and thus value.

Senn, 1987 describes several attributes of information and defines them as the characteristics that are meaningful to the user of each individual item of information. Each individual item of information can be described with respect to accuracy (whether or not the information portrays the situation or status as it is), form (distinction between qualitative and quantitative, numerical and graphic, printed or displayed, summary or detail information), frequency (a measure of how often information is needed, collected or produced), breadth (scope of information, narrow or large), origin (whether information originates from within or outside the organization) and time horizon (whether information is directed toward the past, toward current events or toward future activities or events).

In order to carry out a thorough analysis of the value of information one should be able to put the concept of information as a product into context and look into the attributes that characterise other products (and thus not only the specific attributes of an item of information). Economic goods usually show certain properties such as divisibility, appropriability, scarcity, and decreasing return to use. By contrast, information as a commodity differs from the typical good in that (Glazer, 1993):

- It is not easily divisible or appropriable;

- It is not inherently scarce (although it is often perishable);

- It may not exhibit decreasing return to scale, but often in fact increases in value the more it is used;

- Unlike other commodities which are non-renewable and with few exceptions depletable, information is self regenerative and feeds on itself so that the identification of a new piece of knowledge immediately creates both the demand and conditions for production of subsequent pieces.

These features of information make it difficult to value it as other commodities. In the case of information, value usually means value-in-use or its benefits to the user, i.e. its meaning for the user. The framework described above in Figure 1 can be augmented into the following.

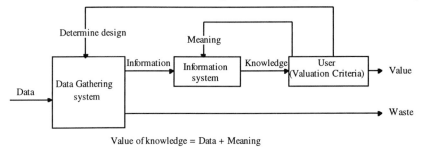

Value of knowledge = Data + Meaning

(Value of product = Material + Function)

Figure 2. Framework for information valuation (adapted from Introna, 1993)

The initial data is put into the data gathering system that structures it. The result can be either waste that is of no use for the current process and can be discarded, or some information that is further input into an information system. An information system as referred to here is any set of people, data, and procedures that work together and that have the capability of storing and delivering information to its users. This system structures, analyses and models the information. The result is a piece of knowledge that can be improved by the user (through his valuation criteria). These user's valuation criteria can also have two effects. One is to help in determining further design of the data gathering system so that it can better refine its output (information). Another possible effect is to give meaning to the output of the information system and thus increase value which is the output of the whole process.

Value of information

There has in recent years been attempts to put a price tag on information and information services in organizations. King and Griffiths (1986) suggested two approaches to estimating the value of information. One is the organization's 'willingness to pay' for information. This is measured by the budget for books, periodicals and other sources of information, plus the individual's investment in time and effort to discover, retrieve and read the information. The second approach is to estimate the cost saving or other advantages that result from the individual having the information. Several studies show substantial gains in having information compared to not having it. The problem is that the amount of data that is needed before useful information can be drawn is usually quite huge. This requires not only a good data gathering system but also good analytical skills for the organization's information system

(Knight, 1990). Furthermore one should keep in mind the old adage 'Garbage In - Garbage Out', i.e. only if the right data is input into the system can it deliver the needed value.

INFORMATION VALUATION IN LOGISTICS

The general view of information given above is that of a product getting increasingly sophisticated and allowing both its providers and its users to acquire more and more gains from it. This viewpoint is highly recognisable in logistics operations where a great many actors on the market are getting more and more information oriented. The operators on the logistics chain are increasingly looking for accurate and speedy information on the status of their shipment. They are more and more trying to render their processes more efficient through the use of information technology. This has opened a whole new business line known as Third Party Logistics where the focus is on managing supply chains by information technology.

Sjöstedt (1994) stated the importance of coupling logistics to information technology. He noted that although substantial advances had been made in creating systems that contain those two domains we have yet to fully acknowledge it. He also noted that we have not fully envisioned the enormous changes to society that could and will result from a dedicated coupling of logistics and information technology principles.

A discussion about information valuation in logistics should take into account the dual aspect of investment versus return. A description of the factors governing investment in information technology is given for instance in Elektroniska affärer, 1995. Other aspects pertaining to the justification of the investment are found in (Triggs, 1993), (Hammant, 1995), etc. We will mostly concentrate on the assessment of the potential returns or gains of the technology.

Information technology dependent gains in logistics

Time gains are often seen as the major benefits of IT. Ordering is done faster, communication is made possible or speedier, handling is done more efficiently, etc. These gains make it very difficult to fairly recognise and value IT investment since they only in very few cases can be priced in economic terms. To be able to determine time gains in logistics operations one needs to define the various components of the transport time from the moment the transport firm takes responsibility of the goods until the moment they deliver it to the customer. Such a description also puts the improvement potential in the logistics time in focus.

A very important corner stone to time gains in logistics from the customer's viewpoint is the contract time. The role of this time is to guarantee that the agreed upon transport time is met i.e. there is no delay on the transport task. In other words the contract time is defined as the last allowable arrival time for a consignment.

Taking reference from the contract time just defined, the actual transport time between two locations or nodes in a network, seen from the shipper's perspective, can be divided into a node time and a link time (see Lumsden, 1995). The link time refers to the time needed for the actual movement of the consignment between the two nodes. The node time is the time needed for various activities at the node location. It can also be divided into two components, namely an *active node time* corresponding to the time when some handling is done on the consignment and a *passive node time* corresponding to the time when the consignment is not subject to any handling at all.

Depending on the transport system used the transport time can allow for a *buffer time* whose role is to guarantee that the contract time constraint is met. The length of this buffer time is very much dependent upon both the design and the efficiency of the transport system used. This leads us to two observations as to where time gains can be accomplished in a transport system via information technology.

Transporting a consignment from one node to another in a network is often done via time tables or some other fixed time constraints. Air cargo that is transported in the belly capacity of passenger flights has to wait for the flight departure time. Seaway transports are also restricted in the same way in that both departure and arrival times are determined in advance. This leaves very little room for improvement in the link time described above, with the help of information technology.

Instead, the node times, both at the sending end and at the receiving end, lend themselves to improvements with better use of information technology. A better receiving procedure at the starting terminal can diminish the time needed for terminal operations. This could for instance be accomplished by notifying the shipper, the customer, when each consignment is received on an individual basis instead of notifying them when the consolidated cargo, e.g. a full ship load, is available. The reduction of the time needed at the sending terminal, which can be called *consolidation time gain* (CTG), can benefit the sending firm in several ways. One is the possibility of having at their disposal more time to ship the goods to the terminal, which could translate into more goods being shipped if necessary. Another benefit of having more time at their disposal is the possibility of having a greater lead-time (in the case of a producing company) which can lead to reduced inventory.

The better receiving procedures just described for the sending end, also hold for the receiving end where the time needed for deconsolidating the consignments can be dramatically reduced through a customer oriented use of information technology. This *deconsolidation time gain* (DTG), depicted in Figure 3, can benefit the receiving firm in several ways. Production can be started sooner if the receiving firm is a producing company. Transportation to another node in the network can be done sooner if the receiving firm is another transportation firm.

In summary the total information technology dependent time gain for a transport operation is the sum of the consolidation time gain which is accomplished at the sending end of the transport task and the deconsolidation time gain which can be accomplished at the receiving end. The concepts just described are all depicted in the following figure.

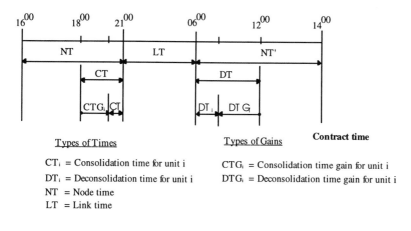

Types of Times

CT$_i$ = Consolidation time for unit i
DT$_i$ = Deconsolidation time for unit i
NT = Node time
LT = Link time

Types of Gains

CTG$_i$ = Consolidation time gain for unit i
DTG$_i$ = Deconsolidation time gain for unit i

Figure 3. Information technology dependent time gains in transport operations

A prerequisite to the gains described above is that information about the shipments be transformed into a physical change, e.g. by devising new loading and unloading systems and strategies. Using information technology to improve transport systems should lead to a positive change in the various components of the transport time, especially at the nodes. But the value of these changes in the components is determined by the customer and is in many cases an ad-hoc value i.e. a value that can change from one situation or user to another.

It was said earlier that the contract time is a guarantee of the agreed upon time. It is thus one of the yard sticks of the customer service levels of the transport firm. Carrying out the transport task within this time frame shows the transport firm's ability to meet its promises. The customer is thus likely to put a value on any improvements of this arrival time.

By the same token improvements at the starting end of the transport task are likely to be valued by the sending firm. The provision for the needed improvements at the starting end of the transport task is determined by how well fitted the transport firm is at transforming its better information about the shipments into a physical system that benefits the sending company.

FROM TIME GAINS TO VALUE OR UTILITY

The time gains described in the previous section can be translated into value, if the logistics system is built in a way that allows the customer to take advantage of the generated extra time. Remember that in Figure 2 information has to be translated into valuable information through some customer valuation criteria. This means that at this stage the customer is an active part in determining what is valuable and what is not. In the case at hand several such customer related values (utilities) can be identified depending on when the customer is notified about his shipment.

In Figure 4 the times needed to perform a loading, transport and unloading task are depicted on the x-axis. The y-axis represents the number of units of load in the current shipment. This number increases in the loading phase until its completion, i.e. until the capacity of the transport resource, e.g. a vehicle, is attained. This is a gradual process which for the sake of simplicity has been represented linearly in the figure. This phase is performed within TL, the total loading time. During the transport phase, which follows the loading phase in the model, there is no change in the actual load. This phase is followed by the unloading phase which is represented by the diminishing number of units until its completion. In Figure 4 the unloading phase is completed within TU, the total time of unloading. In a real world case the loading, transport and unloading phases might not be sequential: there might be a consolidation and/or a deconsolidation of loads within the transport phase. But in this simplified model we discard such specificities.

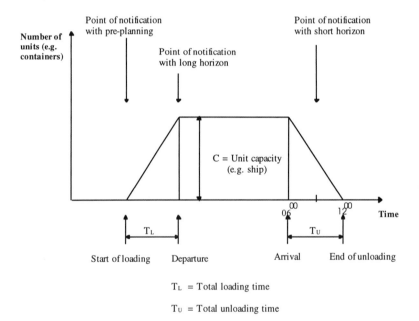

Figure 4. Time gains in relation to information

The first customer related utility is one created by instantaneously getting knowledge of the random order of the items in a shipment when unloading it. This value can be termed deconsolidation time gain with a short horizon, DTGS, due to the fact that the customer is notified at very short notice. The consignments in the shipment need not be in a pre-defined order. In fact they could be in total disorder in the transport resource (vehicle) and still provide for some value. What is necessary is that their status as units (e.g. containers) is reported to the consignee upon arrival in the terminal. The value of this random system lies in that by knowing the status of the consignment the customer is able to determine whether or not the extra time generated by the information system can come to use. An example can be taken from seaway transports. Let us assume that a ship docks at six o'clock in the morning and usually requires six hours to unload. This means that no consignment onboard the ship can be retrieved by the consignee sooner than twelve noon. In this scheme consignments unloaded right after 06 AM will stay at the terminal for six hours or more. Instead a system built on barcodes or transponders for example could be installed, telling the customer exactly when the consignment is unloaded and ready to be carried away. It can with certainty be said that there is a value to this scheme of letting the customers know of the information thus generated. But it is the customer who defines this value through his valuation criteria. This value can differ a lot depending on the type of product transported. Some customers might take full advantage of the time generated by this scheme. Some might find it very difficult to change their plans on such short notice and use the scheduled time instead. But the overriding idea is to let the customer have knowledge of the status of his consignment. A disadvantage of this scheme is that the consignments are not in a given order in the containers upon arrival at the dock and this does not benefit some customers that might be able to take full advantage of the generated time gain. The consignments that were loaded last at the origin and will be unloaded first (in a LIFO scheme) might not be the ones whose owners need to receive as early as possible.

Main deck of Ro-Ro-ship

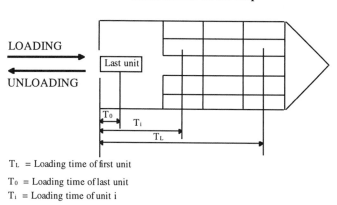

LOADING

UNLOADING

T_L = Loading time of first unit
T_0 = Loading time of last unit
T_i = Loading time of unit i

Figure 5. Sequence for loading and unloading normal box units (e.g. Ro-Ro-ships, containers, trailers)

The second customer related utility can be derived from the one described above by tracing the consignments one link time in advance. Even in this case the consignments are in a random order in the transport resource. But by notifying the customer one link time in advance they will know the random unloading order although it was not planned in advance at the departure. This also allows the customer to have a larger time span for his planning. The utility thus generated is a deconsolidation time gain with long horizon, DTGL, longer than in the first case.

The third and last customer related utility can be accomplished by creating order in the transport unit so that consignments with the greatest potential of economically using the time gains are loaded in a way that allows them to be unloaded first by pre-planning the loading and unloading sequence. This yields a deconsolidation time gain with pre-planning, DTGP. In the LIFO scheme described above such consignments will be loaded last in the containers at the origin, thus allowing for a speedy unloading at destination.

All these utilities will only benefit the customers if they in concert with the transport firm define what their time needs are, thus allowing for prioritisation in the loading and unloading processes.

TWO-DIMENSIONAL GAINS

Taking reference from Figure 4 a trade-off can be discerned between the different time gains and the various points of notification. Each customer could determine which trade-offs fit his operations. Any deconsolidation time gain is traded off with the corresponding point of notification. Notifying the customer as early as possible and generating a high time gain, has a higher value than notifying the customer upon arrival at the destination, i.e. on short notice. This trade-off is the iso-value of the various schemes as described above. The iso-value curve can also have different shapes depending on the kinds of shipments that are referred to.

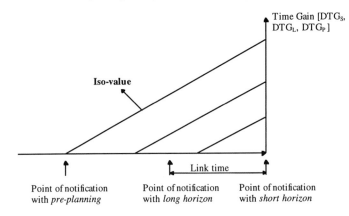

Figure 6. Trade-off between time gain and point of notification

Note that during the whole discussion there has not been any attempt to quantify the value of the different gains. Such a quantification is done by the customer and is very much dependent upon the situation and the needs of the customer.

FINAL REMARKS

The implementation of information technology has up until now focused primarily on the internal efficiency of the transport firm. The continued introduction and application of IT within the transport and the logistics fields require visible improvements for the shipper and his customer. This can only be achieved through abstract changes like time, safety, frequency, etc. Amongst these, the change in time is probably the one most likely to be transformed into value for the customer, the user of the transport system. For this to be realised there is a need for a well thought out definition of the time gain concept. We have attempted in the previous paragraphs to sow the seeds for such a definition. A corollary to this time gain definition is the need that the time gains achieved through the use of IT be transformed into values or utilities for the customer, i.e. the buyer of the transport service. The discussion has focused on how information on the status of a consignment can generate these time gains. The value of this information is gradual: the earlier the information about a consignment is provided to the customer the more valuable it is to him.

REFERENCES

Bonsal P. and Bell M.. (1987). Information Technology Applications in Transport, VNU Science Press

EDI-föreningen Sverige (1995). Elektroniska affärer - Hur värdera nyttan?, EDIS,

Glazer R. (1993). Measuring the Value of Information, IBM Systems Journal, Vol 32, No 1

Hammant J. (1995). Information Technology Trends in Logistics, Logistics Information Management, Vol 8 No 6

Introna L. (1993). The Impact of Information Technology on Logistics, Logistics Information Management, Vol 6 No 2

King D. and Griffiths J. (1986). Measuring the Value of Information and Information Systems, Services and Products, AGARD Conference Proceedings No. 385, January

Knight A. V. (1990). Managing Information, McGraw-Hill International

Lumsden. K. (1990). Identifieringssystem för industri och handel, Studentlitteratur

Lumsden K. (1995). Transportekonomi Logistiska Modeller för resursflöden, Studentlitteratur

Senn J. (1987). Information Systems in Management, Third edition, Wadsworth Publishing Company

Sjöstedt L. (1994) Sustainable Mobility, Chalmers University of Technology, Department of Transportation and Logistics, Meddelande 75

Triggs D. (1993). Justifying Investment in Technology, Logistics Information Management, Vol 6 No 5

10

INFORMATION FLOWS IN LOGISTICS CHANNELS

Thomas Kanflo

ABSTRACT

The concept of creating "Logistics Channels" is frequently suggested in discussions of optimally planning material flows. The channels however, do not only consist of the material flows, it is also of crucial importance that the information flows related to the movement of goods are handled efficiently. The flow of information is different from the flow of material in several ways, the most important perhaps being that information moves not only in the horizontal direction from "left to right", but also horizontally from "right to left" and in vertical directions.

This paper gives a theoretical approach to the concept of channels and presents studies carried out on location with three different companies that are part of their respective logistics channel, focusing on transport and forwarding.

It seems, from the studies, to be so that the forwarder or transporter is having difficulties in setting up efficient information handling systems. Several explanations can be found:

- The transporting companies have normally not been involved in EDI- and related projects between suppliers and manufacturers.

- Especially in international transports there are many parties involved in a transport.

- The transport and forwarding business is very traditional.

- Forwarders are paid to a) arrange a transport from A to B and b) to help in filling in all sorts of documents. The latter is a very important source of income.

The number of information transfer activities decrease with more efficient use of data communication. The opinion of the parties involved, although subjective, clearly indicates that the use of electronic communication increases the efficiency of the logistics channel.

Keywords: Information, flow, logistics, channel, transport, forwarding

BACKGROUND

The concept of creating "Logistics Channels" is frequently suggested in discussions of optimally planning material flows. The channels however, do not only consist of the material flows, it is also of crucial importance that the information flows related to the movement of goods are handled efficiently. The flow of information is different from the flow of material in several ways, the most important perhaps being that information moves not only in the horizontal direction from "left to right", but also horizontally from "right to left" and in vertical directions.

Aim and scope

The aim of this paper is to present the concept of logistics channels. Sections, involving the transporter/forwarder, of three different logistics channels will be analysed from the perspective of efficient handling of information.

Methodology

A theoretical approach to the concept of channel is given. The actual study consists of three different case studies carried out on location with three different companies that are part of their respective logistics channel.

THE CONCEPT OF CHANNELS

The basic idea behind logistics channels

The whole idea of logistics is based upon the thought that an efficient and effective flow of goods and information are two of the most important building blocks of a prospering business. As being successful in business to a large extent means efficiently using your resources, might they be production facilities, personnel, money or any other resource, the logistical ideas and the building of channels should be highly rewarding.

The logistical channel can be looked upon as any other pipe where something is flowing, and analogies can be drawn from other areas: If for instance a river is very wide the water flows slowly downstream, but if the river is narrowed the speed of the water increases. This is a result of a physical law that states that the speed, v, is what you get by dividing the volume flowing in the pipe, V, by the area of the section of the pipe, A. Automatically this means that when the amount of the fluid is constant and the area is reduced, the speed of the flow will increase, see Figure 1.

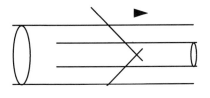

$$v = V / A$$

Figure 1. Reducing channel area means increased speed if the volume shall remain constant

This reasoning can directly be applied to a logistics flow, see Figure 2.

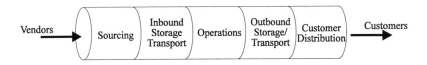

Figure 2. Example of a logistic channel or "pipeline"

Example:

Location A produces pallets of products at a rate of one per day, location B uses them at the same rate. If location A wants to send 10 pallets to location B it can be done in many different ways; all the pallets can be sent as one shipment, they can all be sent as separate shipments, or the lot sizes can be anything in between. From a channel perspective the first alternative is analogical to the wide river above, the 10 pallets flow as one big entity along the chain, arriving at location B at the same time. The average tied-up resources will be five at location A, as the first pallet to be ready for delivery has to wait for the remaining nine; ten during the transport, as they all travel in e.g. the same container; and five at location B, as they are delivered at the same time but used one by one. They also demand a storage space of ten units at both location A and B.

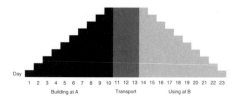

Figure 3. Example 1a

If the channel is made narrower, delivering pallets on a "pull"-command from B, every pallet is shipped as it is ready for delivery and needed by B. The tied-up capital in pallets is reduced dramatically for locations A and B. The average tied-up pallets will be only one half at A and B respectively and three during the transport, but the total number of "transportdays" will be ten. The space needed at A and B will of course be reduced proportionally. On the other hand the frequency of transportation must increase ten times as location B will still need the input goods with the same availability. This means that the resources tied-up in moving the goods, i.e. trucks, will be much larger than before. A balance must be found when calculating the optimal cost of resources.

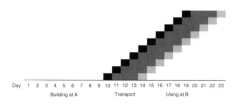

Figure 4. Example 1b

As it is often the parties at locations A and B who initiate changes of this kind, the transporter has to adapt to his customers' demands and produce more frequent transports at a competitive cost. In order to do this he must keep a tight control of the way his resources are used, and one way to do that is to efficiently use various kinds of information; booking information from his customers, information about the goods to be moved, information about the carriers, information facilitating the flow along the actual transport path, and so on.

Different flows in the logistics channel

As discussed in the previous section the logistics channel must include not only the production facility and its operations and internal handling. The movement of goods between sites, both on the supply side and on the distribution side must be included as well. It is also necessary to look at the entire chain from source to consumer, significantly prolonging the channel used in the previous example. This normally means that there are several sets of supplier-transporter-manufacturer-transporter-consumer relations.

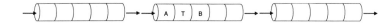

Figure 5. The channel has to connect the initial source to the final consumer

This prolonged channel is considerably more complicated than the previous one and certainly much more complicated than the internal in-factory channel. Unit value of goods and resources as well as their relative value differ a lot at different stages in the channel, creating different demands on frequency, resource capacity and speed, if the utilisation is to be kept at a satisfactory level. Also the different functions of the flow is a source for difficulties. Before production, the goal of the channel is either to just move goods forward in one or several parallel flows, or to match necessary components at the correct time, a converging flow, while when once the product is finished the flow is instead diverging towards the next section of the channel.

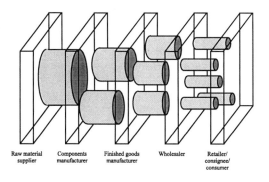

Figure 6. Flow of goods in a section of a logistic channels (Source: IBM, 1993)

Along each physical flow there is always a parallel flow of information holding information about the goods and their present location. As normally several material flows converge into one product, a process that might occur many times before the final product reaches the final consumer, the different information chains are dependent on each other. The primary objective of the information is to create a timely match of components. If the information system fails, the effects on the channel can be serious, causing disorder both down and up the channel (Kanflo, Lindau & Lumsden, 1993). After the final assembly of the product, the goal of the information is no longer to match components, but instead to prepare the consumer in the next step that a product is leaving the channel.

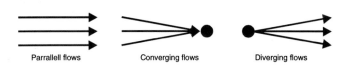

Figure 7. *Different types of (information) flows in logistic channels*

One major problem in this discussion is that as long as the chains are not strictly defined, the manufacturer in each section of the could-be channel considers himself the final consumer and designs information systems that normally consider information flows in only one direction, starting or ending in his operations.

LOGISTIC CHANNEL INFORMATION SYSTEMS

Logistics information

Logistics information is, as has been stated above, very complex. This is particularly true when logistic channels are constructed. Not only does a company have to fulfill its own needs, it also must integrate its systems into several other systems, belonging to other companies, and provide these companies with information that the company itself perhaps does not really consider important. If in addition the company is a part of more than one logistic channel the task of designing these systems might appear almost impossible to accomplish.

Figure 8 illustrates some of the information communicated just to transport a piece of goods from one actor in the channel to another. 18 different transactions are taking place between parties involved. To be complete, however, the internal flow should also be described, as that flow is also quite often a complicated matter. The number of transactions would then be much higher.

As the channel is extended and more segments are added, the complexity of information grows rapidly as more parties want timely information and pre-advice of what is happening to the goods. High information quality is then of utmost importance in the logistic channel (Coyle et al., 1993):

- *Having the right information available*, which means that the true needs for information must be understood;

- *Accuracy of information*, meaning that the information must be not only right but also fresh and without errors;

- *Understandable communication of information*, which means that the one for whom the information is intended, be it a human person or a computer, shall be able to understand the meaning of the information without misinterpretations.

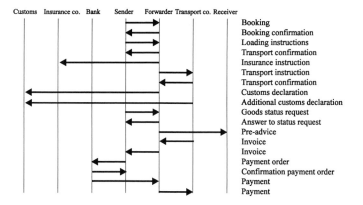

Figure 8. Sub-set of information communicated to send one shipment

Information systems and information complexity in the logistic channel

When companies first started to use computers and information systems in the running of the company, it was generally for bookkeeping and accounting purposes. As the development has continued, the number of applications run on computers has grown at an ever increasing pace. Up until now the focus has however been on internal applications. The first "logistics" applications that came in use were inventory control systems and MRP systems, which were used mainly for internal needs. Now that companies are starting to identify logistics channels as a strategic advantage the information structure of the company is already complicated and designed to work in other ways than would have been preferable from a logistic aspect, i.e. the systems work vertically in the organisation instead of horizontally along the flow. This is a situation that logistics has to accept as it is normally not realistic to start from scratch and build completely new systems. The goals of logistic information system design must therefore be to integrate logistic functions into the existing systems. Where this is not possible or proves too expensive, additional systems or completely new systems have to be built.

As communication of information between sections in the channel is perhaps the most important task for the information systems in the logistic channels, much effort must be put into the design and implementation of communication standards. This is especially important if the company is part of many channels and will be an absolute must if the "virtual corporation" becomes reality, where logistic channels will appear and disappear constantly (Kanflo & Sjöstedt, 1994).

The amounts of data that are communicated in the channels require efficient data communication. Unfortunately an increase in information efficiency also means that the information systems grow more technically complex, and cost more, see Figure 9.

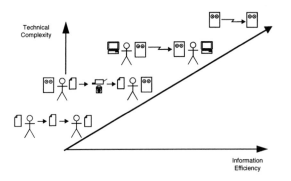

Figure 9. Information efficiency vs technical complexity in four levels of information systems

- The first level of information efficiency is the traditional printed document, forwarded by post or messenger.

- The second level represents the still relatively simple solution of telecommunicated vocal or printed information; the simplicity is, however, a fact only thanks to the standardisation that is the result of years of work by the telecommunication industry.

- The third level is technically more complex than the second; the transmission is basically the same, but interfaces have to be built between the communication equipment and the computer systems, and data transfer formats need to be adjusted to the receiving application if any real efficiency increase is to be expected. A simple file transfer does speed up the movement of data, but it does not necessarily simplify the handling of data.

- The fourth level of efficiency is the first level that should not, if working correctly, need involvement from any humans in the transfer and interpretation of simple data. This is of course technically complex as it requires a very high degree of data transfer reliability and also new programming of applications, allowing automatic data access into, and response actions from, the receiving computer system.

As the standardisation work is constantly progressing in this rather immature area of communication it will within reasonable time not be that difficult to use level three or four applications, though the technical complexity will still be high. If the communication is limited to just a few partners the creation of standards and applications is much simpler than in the case where many partners have to be considered, but the partners are at the same time being isolated from other data communicators. The possibility of being part of more logistics channels is then reduced.

Figure 10. Communication of information in a fully integrated section of a logistics channel

To accomplish a high degree of resource utilisation in the logistics channel, all of the companies involved must be on the third or fourth level of information efficiency. The computer systems have to be able to communicate with one another and for this purpose the use of EDI (Electronic Data Interchange) of some kind is an absolute necessity. In addition to the ability to communicate, the system must of course also be able to produce the necessary orders, plans, measurements, quality reports and so on, needed internally and externally.

Today the information efficiency level in e.g. Sweden is far too low to admit the creation of any totally integrated logistics channels. There are however sections that have been connected, mainly relations between a manufacturer and one or more of his suppliers or customers. Even though many companies know what EDI means, the use of EDI is still very limited. A study made of transport- and forwarding companies shows that only about one fourth of the companies use EDI in one or more applications. Most of these companies use only one EDI application, the Swedish Customs Data System (TDS).

Figure 11. "How deep is your knowledge of EDI?" resp. "Does your company use any EDI applications?" Source: Kanflo & Lumsden, 1994

The belief in EDI is larger than the current use; more than one third of the companies in the study believe that the use of EDI would bring large positive effects to their business and another third thinks that there would be advantages. Only two percent believe that their business would be affected negatively from the installation of EDI applications.

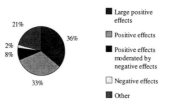

Figure 12. "What effects on productivity and economy would your company get from using EDI applications?"

These figures indicate that even though it may take some time, EDI based communication solutions will eventually spread among companies that are potential members of integrated logistic channels. When that day finally comes the power of well functioning material flows, optimised from a systemwide view of goods and resources, will undoubtedly help logistics become the most important strategic business element and create high system efficiency as well as customer service.

CASE STUDIES

The rest of this paper presents three case studies carried out at different companies, being part of three different logistics channels, or what could be developed into channels. In each studied section three actors are involved, one supplier, one forwarder or transporter and one consumer.

The studies are focusing on the information flows and the tools used for transferring information in every step of the process. Costs of the processes are not considered in this study, although they are naturally of utmost importance when deciding on the most *cost efficient* system, but the assumption, and the subjective opinion of the people involved, is that the *information efficiency* is increased as the use of electronic communication is increased.

In the analysis a model is used where time and goods are moving from left to right. The information flows, normally transferred vertically momentarily between two parties are illustrated either by a straight arrow for information communicated manually or on paper, or by a "flashed" arrow for information communicated electronically.

As actors in the flows, all physical entities involved are included, i.e. not only humans but also computers, fork lifts etc.

Each arrow is numbered. All numbers are explained in the appendixes at the end of the paper.

The warehouse distribution case study

The first case is a part of a logistics channel involving the distribution from an inventory keeping supplier of input material for a producing industry. The general opinion is that the flow of goods and information is very efficient, and has been forced to be so because of the number of variants of input material and the short lead-times demanded by the consumers.

The flow is initiated when the customer sends an electronic order to the supplier. The supplier's main computer automatically communicates a picking order to the warehouse. The order is transferred to a fork lift where it is displayed on the screen of an on-board computer. After picking the goods the picker confirms the pick using the on-board computer and transport documentation is printed on paper. This documentation is given by hand to the truck driver, who after delivery returns the signed documents to the warehousing personnel. A new confirmation is entered into the supplier's central system and an invoice is automatically sent electronically to the customer, who pays by electronic message via his bank.

In this case information is transferred electronically in 9 out of 13 relations. The only times when information is printed or communicated manually is when the transporter is involved. This has partly a practical explanation as the customer has to sign the waybill to confirm the delivery of the goods. The supplier and the customer however consider this a very inefficient way of working, when compared to the other information transfers involved.

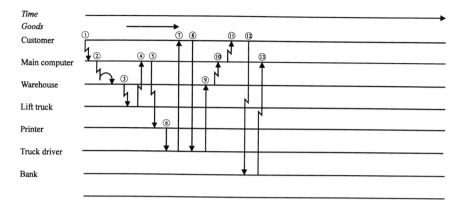

Figure 13. Flow of information in the warehouse distribution channel case

The setup is complicated, but normally functions very well. There are a number of customers using similar setups and the result is:

- Reduced number of personnel for order entry;
- Fewer errors in orders;
- Simplified warehousing administration;
- Reduced lead-times from order to delivery;
- Better integration between warehouse and production at supplier.

The total costs, though no real calculations have been made by any party, are believed to have been reduced.

The JIT import case study

The second case is a part of a logistics channel for Just-In-Time importing of input material from a supplier to a producing industry. The flow of goods and information is considered complex as it has to be matched with other flows from other channels. The manufacturer for this purpose wants electronic notifications of the arrival of the goods to enhance the planning possibilities in his production process.

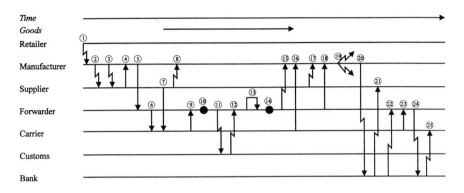

Figure 14. Flow of information in the import channel case

The transport is initiated by the manufacturer on a pull order from a retailer. The manufacturer sends an electronic delivery request as a confirmation of a previously sent forecast. After receiving instruction on loading and unloading a booking is placed vocally with the forwarder. Documents are printed and given to the truck driver who on arrival in Sweden gives them to a forwarder who manually enters them in a system connected to the Swedish TDS. To be able to produce the electronic arrival notice the data is then once again typed into a computer, this time to produce the arrival notice EDI message. On delivery transport the documents are handed over to the manufacturer. The post-transport administration is all done electronically between the supplier and the manufacturer. Between the forwarder and the manufacturer, who pays for the transport, the invoice is however sent by mail.

15 out of 25 information transactions are carried out electronically. Once again the setup is considered efficient between supplier and manufacturer, but not in the relations where the forwarder or transporter is involved. The manufacturer is very worried about the fact that it seems to be so difficult to set up efficient electronic communication whenever there is a transporter involved.

The supplier and the manufacturer have had electronic communication functioning for several years and are convinced it is the most cost efficient way of handling this type of information.

The forwarder export case study

As the first two cases indicated that it was normally relations involving a forwarder or transporter causing inefficient information handling, the third case is focusing on the information flow to, within and from a forwarder. The forwarder studied is a branch in the Göteborg area of one of Sweden's largest forwarding companies. The forwarder in some relationships is part of a more structured logistics channel, but the case studied is a general booking of a consignment going from Sweden to a country in the southern part of Europe.

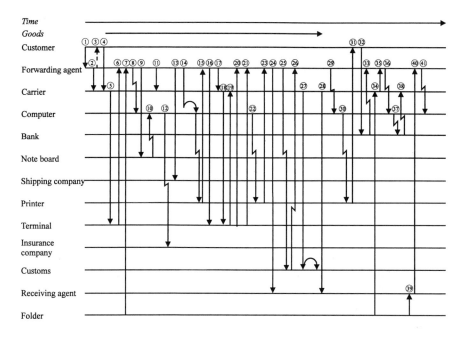

Figure 15. Flow of information in the forwarding export case

The process is initiated when the customer calls to the forwarder to book a consignment. The booking is sometimes followed by a fax (normally the invoice to the receiver of the goods). A preliminary booking is sometimes typed into the computer before the goods are picked up by a collection truck, all arranged by telephone. The complete booking is entered into the system when the freight documentation arrives with the collection truck. A hand-written note is kept for the load-planning process of the international transport. When all shipments to the specific country are collected the forwarder manually plans the loading of the truck going abroad. Insurance information is sent using EDI. A loading list and other documents are printed from the computer system. Space is booked on a ferry. The documents are given to the driver via personnel at the terminal. Signatures are written, a loading confirmation is given and a copy of the loading list is returned to the forwarder. An arrival notice is sent via fax to the agent in the receiving country. Customs data is sent electronically to the Swedish TDS system. The post transport administration is normally done using the computer systems, but invoices etc. are sent by mail.

Only in 17 out of 41 transactions electronic transmission of data is used. The administration is considered very inefficient and unnecessarily complex. Many documents are produced and printed for various purposes, causing a lot of manual distribution of information. Papers sometimes disappear and errors frequently appear as data is entered many times along the way.

CONCLUSIONS

Even though the study is very limited, a few conclusions can be drawn when looking at the flow of information in the logistics channels studied:

Case	Electronic	Manual	Ratio	Subjective efficiency
Distribution	9	4	70%	High
JIT Import	15	10	60%	High (supplier - manufacturer)
				Low (Forwarder)
Export	17	24	40%	Low

It seems to be the forwarder or transporter that is having most difficulties in setting up efficient information handling systems. Several explanations can be found:

- The transporting companies have normally not been involved in EDI- and related projects between suppliers and manufacturers. This is logical as the main relation is a product seller-buyer relation where the transport is not generally discussed. It is however unlucky as much of the information related to the product, and especially its movement, involves or is even created by, the transporter.

- Especially in international transports there are many parties involved. In foreign countries forwarders often use agents, meaning that e.g. the European transport market is very complex with many unstable and loose relations, not encouraging often tight EDI relations, as the open standards have until now not really been there.

- The transport and forwarding business is a very traditional one. The focus has for a long time been on doing things "in the old way", using computers mainly as electronic typewriters. In recent years however, things have begun to change. Forwarders today see themselves more as logistics providers that can really play an active role in the changing industrial processes.

- Forwarders are paid to a) arrange a transport from A to B and b) to help filling in all sorts of documents. The latter is a very important source of income, especially when the pressure is tough on lowering the actual transport rates. This means that it is likely that the will to adopt new technologies reducing the number of documents sometimes has been lacking.

The number of information transfer activities decreases with more efficient use of data communication. One reason could be that the companies, when doing projects in the area, at the same time question the necessity of all the steps in the chain, and as a consequence eliminate unnecessary information movement. Another important reason is of course that sending information electronically takes away the need to send all information to printers, saving not only a lot of paper but also a lot of trouble, as printers can be very problematic, can run out of paper and all sorts of other things.

The opinion of the parties involved, although subjective, clearly indicates that the use of electronic communication increases the efficiency of the logistics channel. As it is often the intuitive opinion of people that set the course of a company this means that the observed development will continue, and that the importance of efficient information systems in the logistics channels will increase. It also means that the work to find methods of really judging the efficiency of information systems must be further stressed, so that the subjective opinions can be confirmed or rejected using objective investment analysis.

REFERENCES

Coyle, Bardi and Langley, *The management of Business Logistics*, West Publishing Company, St paul, USA, 1992.

APPENDIX A

The information relations in the *warehouse distribution* case study:

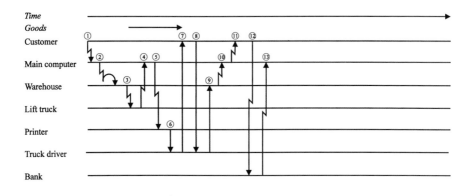

1 Customers place their orders directly into the main computer of the wholesaler (*ordering by phone or telefax to an incoming orders department is also available*).

2 The main computer directs the orders to the correct warehouse, where the orders, still in the computer, are sorted and given to the right order picker.

3 The picking "list" is then sent to the on-board terminal on the lift truck.

4 The picker confirms his picks directly on the terminal, and the confirmation goes into the main computer.

5 When the order is complete a consignment note is printed in the warehouse.

6 *The consignment note is given to the driver of the distribution truck.*

7 *The driver gives the note to the customer for signature.*

8 *The customer gives the signed note back to the driver.*

9 *The driver returns the consignment note to the warehouse when he gets back.*

10 The delivery is confirmed on the warehouse terminal.

11 The main computer prepares the invoice and sends it electronically to the customer (*invoicing by mail is also available*).

12 The customer instructs the bank to pay the invoice.

13 The bank transfers the payment to the receiver.

APPENDIX B

The information relations in the *JIT import* case study:

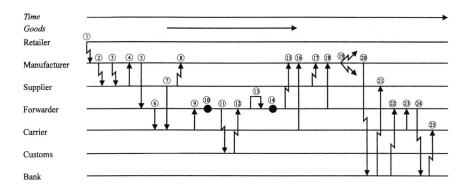

- The transport is initiated by the manufacturer, who also pays for the transport.

- The goods are picked up at the supplier and do not pass through a terminal in the exporting country.

1 Retailers place orders from customers into the manufacturers computer.

2 The manufacturing company sends forecasts to their supplier electronically via ODETTE.

3 Calloffs are sent from the manufacturer to the supplier every two weeks, also via ODETTE.

4 *The supplier phones back to the manufacturer with instructions on loading and collection.*

5 *The manufacturer phones a transport booking to the forwarding agent.*

6 The forwarding agent phones instructions to his carrier.

7 *The supplier hands over transport documents to the carrier that collects the goods. The supplier has put an ODETTE flag on the package.*

8 The supplier sends a pre-advice message to the manufacturer via ODETTE.

9 *When the carrier reaches the forwarding agent in the receiving country he hands the transport documents over to the forwarder.*

10 *The forwarder types the needed data in to the Swedish Customs System, TDS.*

11 The forwarder sends the message to the customs via TDS.

12 The TDS automatically sends a reply to the forwarder.

13 *The forwarder receives at this time or earlier a pre-advice from his office in the exporting country. The pre-advice is sent by telefax.*

14 *The forwarder types data needed in to a terminal in order to generate an ODETTE message.*

15 The pre-advice ODETTE message is sent from the forwarding agent to the manufacturer.

16 The carrier hands over goods and transport documents to the manufacturer. The manufacturer matches the ODETTE flag on the package (7) with the pre-advice from the supplier (8).

17 The supplier sends an ODETTE invoice to the manufacturer.

18 The forwarder sends an invoice via mail to the manufacturer.

19 The manufacturer sends copies of the ODETTE invoice or *types in to the computer system and sends copies of the manual invoice.*

20 The manufacturer sends payment instructions to the bank using tape sent by mail.

21 The bank transfers money to the supplier.

22 The bank transfers money to the forwarding agent.

23 The carrier sends an invoice to the forwarding agent via mail.

24 The forwarding agent sends payment instructions to the bank using tape sent by mail.

25 The bank transfers money to the carrier.

APPENDIX C

The information relations in the *forwarder export* case study:

1 *The forwarding agent receives a booking by phone from a customer, sometimes followed by a telefax.*

2 *The forwarding agent instructs his carrier on loading and collection, by phone.*

3 *If it is not a regular transport, the forwarding agent phones the customer to prepare him for the collection.*

4 *The customer hands transport documents to the carrier.*

5 *When the collecting carrier reaches the terminal he passes the documents over to the terminal personnel.*

6 *From the terminal the documents are sent or delivered to the forwarder.*

7 *The forwarder uses a printed quotation to prepare for the charging of the transport.*

8 *The booking is typed into the computerised forwarding system.*

9 *The initial booking note is saved on a noteboard for loadplanning.*

10 When all the bookings for a certain transport have arrived the bookings are tied to a lorry or carrier in the system.

11 *The forwarder phones the carrier and instructs him on collection and loading.*

12 Insurance instructions to the insurance company via EDI.

13 The forwarder books space on the ferry, by phone or telefax.

14 The forwarder prepares the loading list in the system and prints it.

15 The forwarder walks to the printer and collects the loading list.

16 The loading list is delivered to the terminal by pneumatic dispatch or messenger.

17 The carrier collects the transport documents from the forwarder.

18 The carrier hands documents over to the terminal personnel for signature.

19 The signed documents are returned.

20 The terminal personnel phones the forwarder to say that the loading is complete.

21 The signed loading list is returned to the forwarder.

22 A pre-advice for the agent is printed.

23 The forwarder walks to the printer and collects the pre-advice.

24 The forwarder sends the pre-advice to the agent via telefax.

25 The export is reported to the customs using the Swedish Customs System, TDS.

26 Reply from the TDS.

27 The carrier hands the export documentation over to the customs agents in the importing country and returns it to the Swedish customs by mail.

28 The remaining documentation is delivered by the carrier to the receiving agent.

29 The forwarder prepares the invoice in the computer.

30 The invoice is printed.

31 The invoice is sent to the customer by mail.

32 The customer instructs his bank to pay the invoice, by *mail* or computer (tape).

33 The payment is transferred to the forwarding agent.

34 The carrier prepares the invoice to the forwarding agent, studying the written agreement.

35 The carrier sends the invoice to the forwarding agent via mail.

36 The forwarder enters the value into the computer system.

37 The bank is instructed to pay the invoice, by tape.

38 The payment is transferred to the carrier.

39 The receiving agent prepares the settlement claims from the exporting forwarding agent, using the written agreement.

40 The receiving agent sends his claims to the forwarding agent for settlement, by mail.

41 The claims are entered into the computer system on the credit side of the receiving agents account.

11

USING DEA FOR DECISION SUPPORT AT COMBINED TRANSPORT TERMINALS

Stefan Sjögren

ABSTRACT

Studies of road-rail terminals show that the terminals are operated on different efficiency levels. In a combined system with unit prices for terminal handling and lack of competition among the terminals the incentive for efficiency improvements has hitherto been small. A decision support system may contain bench-marking strategies, cost information, and demand expectation. This study will illustrate how performance indicators, constructed by using DEA (Data Envelopment Analysis), can be used for internal bench marking in a monopolistic type of operation. The shaping of such a decision support would, then, contribute to actions towards a higher efficiency as each terminal may compare its results with a set of reference terminals.

Keywords: DEA, decision support, combined transport terminal, efficiency

INTRODUCTION

"Intermodal terminals are crucial points of transport logistic chains. Improvement in the quality of the terminal operations is considered as a key issue for the optimisation of the overall quality of the intermodal transport system." (EU White Paper, 1994)

A terminal is a point in a transport system where the transport is brought to a temporary stop. A stop means increased costs. Therefore, the stop must, at the same time, add some kind of value. There are terminals where this increase of value consists of consolidation or breaking of the goods in order for them to be transported in optimum volumes. Examples are break bulk terminals, railway stations and airports. The terminals can also be used for storage of goods. A terminal which is a part of a combined transport system is different from such terminals. The increase of value which this type of terminal can add is the transfer of the load unit between two transport modes. The idea is that each specific transport mode shall be used where it will be most advantageous. Rail should be used for long hauls and road transport for short distances, where flexibility is demanded.

The terminals studied are important parts of a combined transport chain and represent the points where reloading from road transport to rail transport takes place. Combined transports are defined as follows: The goods shall travel in load units from the place of dispatch to the place of delivery. Load units are trailers, swap bodies, containers. The load unit must change transport modes at least once from the place of dispatch to the place of delivery.

The above definition indicates three essential requirements which have to be complied with in order for a haulage of goods to be called a combined transport. One of them is that the load unit must change transport modes. Possible combinations are for instance road/rail and rail/sea. In this study only the combination road/rail has been considered.

The terminals studied are especially adapted to the moving of load units between road and rail. They have a limited area of railway tracks for trains and space for trucks. At a terminal there is also equipment for handling, in most cases cranes or lifting trucks, for lifting the load units. The terminals are generally located adjacent to marshalling yards. They can, then, utilise the existing capacity for shunting of trains.

The location of the terminals depends on the potential amount of goods suitable for combined transports within different areas. Due to the lack of competition among terminal operators it has not been necessary to take other factors into consideration. High initial investment costs have been motivated by a high amount of potential goods. The high investment costs have been entrance barriers for other potential terminal operators.

One way to minimise the negative effects of a monopolistic situation, where incentives for efficiency improvements can be smaller than they are in a competitive market, would be to increase the internal competition. This could be called an "artificial" competition. In Sweden, costs and incomes are compared between the terminals, on an overhead basis. The results from this comparison are distributed to terminals. This information will, then, help terminal staff to reveal and increase their efficiency. The information consists mostly of key-ratios. The purpose of this study is to develop a way of using DEA as a method to reveal inefficiencies, and to distribute these results among the terminals. The terminals will, then, have incentives for increasing efficiency. One difference between a monopolistic situation and a free competitive situation is access to information. Information in a free competitive situation is said to be free and obtainable for everyone. A terminal with its high investment costs, that nearly makes it a natural monopoly, can thereby create an efficiency incentive information system. The main purpose of this study is to show how information about efficiency improvements can be created, evaluated and then distributed among the terminals. This will be carried out in three steps.

Step 1. *Development of a method to measure efficiency at terminals.*

Step 2. *Comparison of specific terminals with reference units and the best in the branch.*

Step 3. *Studies on terminals showing how different "benchmarks" can be realised.*

STEP 1 - TO DEVELOP A METHOD IN ORDER TO MEASURE EFFICIENCY AT TERMINALS

In step 1, the method for measuring efficiency will be introduced. The data collected from the Swedish and the German terminals shall be adapted to this method by operationalisation of the activities taking place at the terminals into measurable variables. The results from the efficiency measurements will also be presented briefly.

The DEA-method

A non-parametric method called DEA (Data Envelopment Analysis) will be used. This method was developed by Charnes, Cooper & Rhodes (1978) for measuring efficiency in "non-profit centres" or Decision Making Units (DMUs). "Non-profit centres" or DMUs are characterised by the difficulties in finding a production function. In a study by Sjögren (1996), data on combined transport terminals was collected to test the method's possibilities in the area of transport services. The adaptation of the method to combined transport terminals will be described in brief here.

DEA was originally developed for use in non-profit organisations but the fields of applications have increased. For analysis of non-profit organisations see for example Greenberg & Nunamaker (1987) or Banker, Conrad & Strauss (1986) who compare DEA with more widely used ratio measures. Prior & Sola (1994) use DEA for estimating hospital performance. Jennergren & Obel (1986) use DEA for analysing research performance for several research institutes in Denmark. For analysis in private sectors see e.g. Hjalmarsson, Bjurek & Isaksson (1991), Bergendahl (1993), analysing DMUs in banking and insurance, Försund (1992) analysing ferry transports in Norway, and Andersson (1996) for analysing restaurants. The application of DEA within the transportation sector has hitherto been modest. Some examples are Baker (1989), Gathon & Perelman (1992), Kumbahkar (1989) and Tulkens (1992).

Unlike the classic econometric approaches that require a pre-specification of a parametric function and several implicit or explicit assumptions about the production function, DEA requires only an assumption of convexity of the production possibility set and uses only empirical data to determine the unknown best practice frontier.

DEA is a suitable method for analysing efficiency and productivity in multiple input - multiple output production. Instead of specifying an analytical production function, a production possibility set is measured by observations from the DMUs (Decision Making Units). The production possibility set is enveloped with linear programming and forms a convex front. This convex front, called the "best practice front", is composed of plan areas on which the most efficient DMUs are located. Depending on the demarcations of the front and by comparing different reference points different efficiency measures can be obtained.

To explain the idea of DEA, we will start from classic production theory following Farrell (1957). An "efficient production function" is used for measuring an efficiency that takes multiple input and multiple output into account. An "efficient production function" can be described as the output that a perfectly efficient firm could obtain from any given combination of input.

Under the assumption of constant returns to scale it is possible to represent a "simple" isoquant diagram and discuss the different efficiency measures. Farrell distinguishes between technical, price (or allocative), overall and structural efficiency.

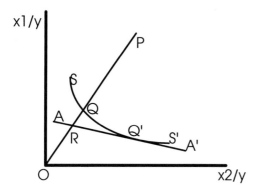

Figure 1. Efficiency measures in a simple case. Source: Farrell (1957)

Technical efficiency is defined as the ratio OQ/OP. This means that a firm operating at Q uses less input (x_i) than a firm operating at P when producing the same amount of output (y). The isoquant SS' represents the various combinations of the two factors that a perfectly efficient firm might use to produce the same amount of output. Price, or allocative, efficiency is defined as a measure that considers the use of the various factors of production in the best proportion, in view of their prices. AA' has a slope equal to the ratio of the price of the two factors. The cost of production at Q' will be less than that at Q. Price efficiency of a firm operating at Q is, then, the ratio OR/OQ.

The total ratio OR/OP is the overall efficiency of a firm if it is to be perfectly efficient both technically and in respect of prices. Overall efficiency is the ratio of minimal costs at the frontier and observed costs of producing the observed output at observed input prices.

Farrell emphasises the difficulties in assuming an efficient production function theoretically. "Although it is a reasonable and perhaps the best concept for the efficiency of a single production process, there are considerable objections to its application to anything so complex as a typical manufacturing firm, let alone an industry" Farrell (1957:255). Accordingly, there are some problems with reference to specifying a theoretical production function when multiple input gives multiple output. Instead the measure of efficiency should be based on the best results observed in practice. That is estimating an efficient production function from observations of the inputs and outputs in a number of firms. Each firm is represented by a point in an isoquant diagram. The point can, then, be located on, or north-east of, the isoquant SS'.

The curve SS' constitutes the estimate of the efficient isoquant. The technical efficiency of the firm is obtained when comparing it with a hypothetical firm that uses the factors in the same proportions. "The hypothetical firm is constructed as a weighted average of two observed firms, in the sense that each of its inputs and outputs is the same weighted average of those of the observed firm" Farrell (1957:256). Farrell emphasises that this is the essence of the method and the idea is not only to present the observations in an isoquant diagram.

Measuring efficiency with DEA makes it possible to create a reference technology (the best practice front) on basis of empirical observed input-output combinations. The method will also determine the efficiency of the production units compared with the reference technology. The creation of a reference technology presumes the introduction of the assumption of returns to scale. Increasing, constant or decreasing returns to scale indicate that an additional input gives increasing, constant or decreasing average output per unit input.

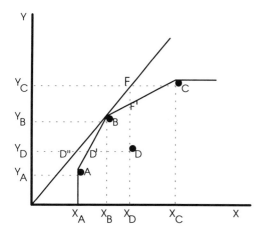

Figure 2. Construction of a reference technology under the assumptions of returns to scale

In Figure 2 there are four possible input(x)-output(y) combinations, A, B, C and D. The different convex fronts that envelop the observations specify different choices of reference technology (returns to scale). XA-A-B-C defines a reference technology characterised by variable returns to scale (VRS). O-B-C defines a reference technology characterised first by constant and then by decreasing returns to scale. This is called decreasing returns to scale

(DRS). O-B-F defines constant returns to scale (CRS). The figure gives rise to the following three notations:

1. If two input-output combinations are technically possible then all weighted average combinations of these two combinations are also technically possible.

2. The reference technology defines the smallest number of input-output combinations that contain the empirically observed combinations, which corresponds to the assumption of returns to scale.

3. The possibility space (the area to the right of the front O-B-F in Figure 4) for CRS technology embraces the possibility space for the DRS technology which accordingly, embraces the possibility space for the VRS technology (X_A-A-B-C). The CRS technology is the least restrictive.

The unit D will be used to illustrate the measure of efficiency. In the case of VRS the following measures can be calculated, (e.g. Hjalmarsson et al, 1991):

$E_1 = Y_DD'/Y_DD$ (decreasing in input)

$E_2 = X_DD/X_DF'$ (increasing in output)

$E_3 = Y_DD''/Y_DD = X_DD/X_DF$ (scale efficiency, total)

$E_4 = E_3/E_1 = Y_DD''/Y_DD'$ (pure scale efficiency, input adjusted)

$E_5 = E_3/E_2 = X_DF'/X_DF$ (pure scale efficiency, output adjusted)

For CRS the efficiency measure E1 equals E2 which equals E3.

The DEA-method estimates Farrell's measures of efficiency under alternative assumptions of returns to scale. These estimations are made by solving a linear programming problem for every single production unit. The following linear programming problem seeks the measure E1 above. When changing the DEA-model from assumptions of constant returns to scale, to variable returns to scale, the data collected will be enveloped more narrowly. The difference in efficiency score can be called disadvantage of scale.

Mathematical description of DEA

Let N be the number of production units, producing r outputs using m inputs. Let observed input for an arbitrary unit g be:

$$X_g = (x_{1g},...,x_{mg}) \qquad (1)$$

and observed output:

$$Y_g = (y_{1g},...,y_{rg}) \qquad (2)$$

The E_1 efficiency measure for unit g is determined by solving the following linear programming problem:

$$Min\ E_{1g} \tag{3}$$

$s.t.$

$$E_{1_g} x_{i_g} \geq \sum_{j=1}^{N} \lambda_j x_{i,j} \ (i = 1,...m) \tag{4}$$

$$Y_{k_g} \leq \sum_{j=1}^{N} \lambda_j y_{k,j} \ (k = 1,...,r) \tag{5}$$

$$\sum_{j=1}^{N} \lambda_j \geq 0 \tag{6}$$

(6) is the constraint that determines the reference technologies. The above formulation is under assumptions of constant returns to scale. For variable returns to scale the constraint (6) is changed to:

$$\sum_{j=1}^{N} \lambda_j = 1$$

The DEA-method must also be adapted to the operations taking place at the terminals. This adaptation is made by modelling the terminal and thereby operationalising the activities into input variables and the production into output variables.

In this study, we will not focus on cost minimisation. Instead we will look at relative technical efficiency for the terminals. One reason for this is the lack of information about costs, and the occurrence of subsidies influencing the market prices. Another reason is that the resource utilisation will be viewed and if possible reduced, independently of costs that often are dubious in interpretation. The Ei measurement refers to technical efficiency according to Figure 1.

Operationalising the activities into variables and measuring the efficiency of terminals

Different activities are carried out at a terminal. These activities give rise to resource utilisation. In the first column of Figure 3, the activities are presented. To perform these activities, resources are used (presented in column 2). The third column consists of variables that have been chosen to describe the resource utilisation.

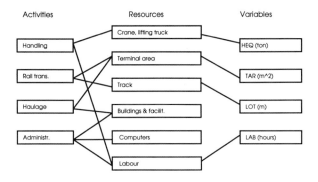

Figure 3. Activities, resources, and variables used to describe a terminal

The main reason for making the operationalisation in Figure 3 is to facilitate data collecting and processing. This makes it possible to choose between different measurement units. In Table 1, the DEA results of terminal efficiency are shown. Four models are created with varying input and output variables, as well as varying assumptions to scale (for more details, see Sjögren 1996). The following input variables are used:

- Labour hours (LAB);

- Total handling capacity (HEQ);

- Length of track (LOT);

- Area of the terminal (TAR).

The above input variables are supposed to reflect the resource utilisation and the activities which have taken place within the terminal. The service being produced at the terminal is handling of load units. In our models the output is measured with the output variable ENT (total handling per year) The model Ec is under assumption of constant returns to scale. The model Ev is under assumption of variable returns to scale.

The data collected comes from 51 terminals. 13 of the terminals are located in Sweden. The other 38 terminals are located in Germany. The terminals in Sweden and in Germany use the same type of resources and produce the same products and a comparison of efficiency is then possible to make. Only small adjustments in the data have been made. None of the Swedish terminals has a 100% relative efficiency measured with model Ec. 10 German terminals turned out to be efficient. In model Ev, two Swedish and 15 German terminals are efficient. There is a large difference between the peers and the terminals with the lowest scores. The average efficiency scores, shown in Table 1, are higher for the German terminals than for the Swedish ones.

Table 1. Average efficiency scores for models Ec, and Ev

	Terminals	Mean (S_0)	Standard deviation
	All	64.60	26.84
Ec	German	68.69	28.32
	Swedish	52.63	17.89
	All	72.26	24.61
Ev	German	77.19	24.52
	Swedish	61.79	21.86

The measurements show that the German terminals are more efficient on average than the Swedish terminals irrespective of model. There is a large dispersion between the efficient and the inefficient terminals. This indicates that there is a potential to achieve increases of efficiency. The large differences in efficiency, of course, depend on the large differences in utilisation of resources.

Classifying the efficiency scores according to total output shows us that large terminals (more than 50 000 load units per year) are more efficient on average than smaller ones.

Table 2. Average efficiency scores classified according to total output

Terminals	Average	Standard deviation
Less than 20 000 units/year	52.00	32.69
Between 20 000 and 50 000 units/year	68.08	21.80
More than 50 000 units/year	90.40	13.38

Six of the peers (relatively efficient terminals) are handling more than 50 000 load units per year. Three of the peers belong to the smallest group. This proves that it is possible for a smaller terminal to be 100 % relatively efficient.

The large dispersion in efficiency measures shows that there are possibilities for the terminals to increase their efficiency. Terminals operating on the wrong scale is one source of inefficiency; most of the inefficiency, however, is of a technical nature. Instead of increasing all allocations of resources proportionally, the object is to change the input/output mix and to restructure the resource factors. This is evident when comparing measures of efficiency with constant returns to scale (Ec) and measures of efficiency with variable returns to scale (Ev) (e.g. Banker, 1984). Average efficiency for the former is about 65% and for the latter about 73%. 8% of the total inefficiency can be explained by scale advantages (73% - 65%). 27% (100% - 73%) depend on "pure" technical efficiency. Both models have a large dispersion. DEA is used to show inefficiencies and at the same time indicate possibilities of increases of efficiency. When different assumptions about returns to scale are introduced, it will be decided implicitly how inefficiencies should be interpreted. If constant returns to scale are assumed, this means that the terminals that do not lie on the production front are technically inefficient. They can, then, change the mix of resources or use them more economically and thus increase efficiency. When variable returns to scale are assumed, this means that part of the measured inefficiency depends on the fact that the terminals operate on a wrong scale.

Another analysis, not presented here, shows that the terminals are, relatively speaking, better at adapting LAB than other variables to the number of produced units. This could be explained by the fact that this variable is the easiest one to influence on a short run. Large improvements in efficiency can be achieved if the utilisation of the variable TAR could be decreased for the inefficient terminals. To a large extent this is also true for the variables length of track, LOT, and handling equipment, HEQ (Sjögren 1996).

STEP 2 - COMPARISON OF SPECIFIC TERMINALS WITH REFERENCE UNITS AND THE BEST IN THE BRANCH

In step 2 each terminal will be compared with the best terminal or with a reference terminal. The results from the DEA measurements point out that large improvements in efficiency can be made. However, they do not reveal how these improvements could be obtained. In step 2 it will be shown how a specific terminal can be compared with other efficient terminals and a reference terminal. The reference terminal and the efficient terminals constitute benchmarks to specify efficiency goals. This course of action can give a more detailed analysis of efficiency. This type of analysis attempts to find the causes of obtained efficiency as well as indicate to what extent efficiency improvement can be made. In the analysis each variable's contribution to the efficiency score is investigated. This can be accomplished in two ways. One way is to compare the examined terminal with the efficient terminal. Another way is to compare the examined terminal with its reference terminal.

Comparing the terminals with reference terminals

A reference terminal is a fictive terminal created by a linear combination of efficient terminals. This will give guidance regarding the direction in which the inefficient terminals should move in order to improve their efficiency. The chosen terminals are S6, S7, S11, T9, T10 and T40. Terminals T9 and T40 are efficient terminals. The choice of terminals to be investigated was based upon different factors as location, received efficiency score, number of units handled and handling equipment used.

Table 3. Comparison with efficient terminals

Terminals	S6			S7		S11		T10			
Eff. term.*		T9	T40		T40		T40		T9	T15	T40
Lambda		0.443	2.243		0.893		2.504		0.110	0.164	3.430
Scale factor		1.375	5.600		2.193		3.509		0.947	1.154	5.433
LAB	53992	35796	27333	11673	10704	29632	17126	29286	24650	18768	26520
HEQ	168	107	168	87	66	139	105	163	74	88	163
LOT	1800	1800	1279	500	500	900	800	1460	1239	1297	1239
TAR	100000	43973	19600	28000	7675	45000	12281	30280	30280	30280	19017
ENT	69095	103447	89365	14249	34996	39966	55993	72118	71234	64279	86705

* The efficient terminal's unscaled inputs and outputs are shown in Table 4 for terminals T9 and T40. T15 is a terminal that is included in the reference terminal for T10.

In Table 3 each reference terminal is created by efficient terminals shown for each terminal. Terminal S6, for example, is measured against a linear combination of T9 and T40. The lambda value consists of the share of each efficient terminal being a part of the linear combination which creates the reference unit. The scale factor shows the appearance of each efficient terminal when adjusted to a level comparable to the examined terminal. The scale factor is determined by the lowest ratio of each input respectively. The efficient terminals of S6, T9 and T40, both produce more output using less resources.

Terminals S7 and S11 are compared with the efficient terminal T40. The scale factor shows how many more times outputs S7 and S11 can produce with their resources in order to be as efficient as T40. In the reference terminal for T10, three efficient terminals are included. An efficient T10 can consequently be created by taking 0.11 times T9, 0.164 times T15 and 3.43 times T40.

Table 3 can be used to show where efficiency improvement ought to be made for each examined terminal. This procedure can be used as a decision support for the terminals. It should be pointed out that each examined terminal is compared with existing terminals. In Table 4 the information available when comparing each examined terminal with its reference terminal is presented. The reference terminal in the table demonstrates how few resources every examined terminal needs to be 100% efficient and to be on the efficient front (cf Figure 2). The reference terminal is said to constitute the efficiency goal for the examined terminals.

Table 4. Comparison between examined terminals and reference terminals

		LAB	HEQ	LOT	TAR	ENT
	Value of variables	53992	168	1800	100000	69095
S6	Reference terminal	22472	102	1090	22007	69095
	Possible increase in eff.	58.4	39.4	39.4	78.0	0.0
	Value of variables	11673	87	500	28000	14249
S7	Reference terminal	4358	27	204	3125	14249
	Possible increase in eff.	62.7	69.2	59.3	88.8	0.0
	Value of variables	29632	139	900	45000	39966
S11	Reference terminal	12224	75	571	8766	39966
	Possible increase in eff.	58.7	45.9	28.6	80.5	0.0
	Value of variables	26032	78	1309	31978	75229
T9	Reference terminal	26032	78	1309	31978	75229
	Possible increase in eff.	0.0	0.0	0.0	0.0	0.0
	Value of variables	29286	163	1460	30280	72118
T10	Reference terminal	22262	124	1110	19816	72118
	Possible increase in eff.	24.0	24.0	24.0	34.6	0.0
	Value of variables	4881	30	228	3500	15958
T40	Reference terminal	4881	30	228	3500	15958
	Possible increase in eff.	0.0	0.0	0.0	0.0	0.0

If the "possible increase of efficiency" differs among the variables (see for example S6) it is a proof of slack. A slack means that a reduction in use of this resource, ceteris paribus, will not affect the efficiency score. For S6 there is a slack received for the variables LAB and TAR. The table gives a good overview over efficiency goals for each terminal and indicates where resource efficiency improvements would be most effective.

STEP 3 - QUALITATIVE STUDIES OF THE TERMINALS IN ORDER TO EXAMINE POSSIBILITIES FOR IMPROVEMENTS OF EFFICIENCY

In step 1, DEA was used to measure efficiency at terminals. In step 2, procedures for analysing the DEA results in detail for each terminal were described. In these two steps, a lot of quantitative data was produced, the intention being that the quantitative data should be distributed among the terminals to support decisions concerning the terminals' investments, operations and resource utilisation. Ideas for using DEA as a benchmarking strategy can be found in Parsons (1994) and in Fried, Lovell & Eeckaut (1992). In step 3 we will see if these quantitative data can be realised. Case studies on chosen terminals will, then, reveal possibilities for decreasing the resource utilisation and/or increasing the output. The benchmarks, consisting of efficiency terminals and constructed reference terminals, are not possible to put into practice in full. All the quantitative data that each terminal receives must be put against the terminal's economical, physical and environmental situation. In this step, results from qualitative research on six chosen terminals are presented (for more details, see Sjögren 1996). The aim of this qualitative study is to show if terminals can improve their efficiency to such an extent as the quantitative data suggests. This procedure will strengthen the decision support. The study focuses on the chosen variables describing the resource utilisation at each terminal.

Terminal structure

The terminals fall, roughly, into two categories. The first is made up of typical crane terminals where the trucks drive in lanes under the crane tracks and where most of the load units are deposited. The other type of terminal uses lifting trucks or both lifting trucks and cranes. The trucks move freely on the grounds. This type of terminal does not use the space just as efficiently and has, in most cases, more tracks. The six terminals studied use different types and combinations of handling equipment for lifting the load units. The Swedish terminals use both lifting trucks and cranes and the three German terminals use, to a larger extent, either only lifting truck or only crane. At three of the terminals included in the case studies, the handling equipment was regarded as a scarce resource. At the other terminals the handling equipment was never used to its full capacity. The reason for using both lifting trucks and cranes was a wish to be as flexible as possible and not having to commit oneself to the use of one technique only. The cranes available could not handle the heavy trailers which are allowed and, therefore, it was necessary to invest in lifting trucks. One crane could not cover all tracks. The crane was considered to have the advantage of being faster for the lifting itself of the load units. The advantages of a lifting truck are its flexibility and its ability to deposit load units over large areas. Terminals with crane only, show a somewhat higher degree of efficiency.

Handling equipment

The examined terminals use different kinds of handling equipment in various combinations. T9 and T40 use only cranes. Terminal T 10 uses only lifting truck. The Swedish terminals S6, S7 and S11 use both lifting truck and crane. The persons interviewed at terminals T9, T40 and S7 all agreed that the handling equipment often is a scarce resource. At terminal T10 they had a spare lifting truck. At this terminal and at terminals S6 and S11 the handling equipment never was regarded as a scarce resource. Using both crane and lifting truck gives a high degree of flexibility and the problem of using only one type of technology is avoided. The cranes do not always have capacity to lift trailers weighing over 37 ton. At terminal T10 the reason for using only lifting truck is that a crane is not able to cover all the tracks which are both numerous and short.

The advantage of the crane was its capacity for fast lifting of each handling unit, though, mostly, two men were required to operate the crane. The advantage of the lifting truck was its flexibility and possibilities of depositing the handling units over a large area.

The terminals using only cranes received 100% efficiency. In Table 5, the terminals are classified into three groups. The average DEA efficiency scores for each group are also presented. A slight difference in received efficiency score is noticed.

Table 5. Average efficiency score of the terminals when classified according to the use of handling equipment (all 51 terminals)

	Average (Ec)	Standard deviation
Terminals using only cranes	68.08	33.62
Terminals using only lifting trucks	64.42	25.04
Terminals using both cranes and lifting trucks	61.50	20.36

The terminals using only cranes are showing a slightly higher efficiency on average. The fact that 7 out of 10 efficient terminals only use cranes emphasises the results above. The terminals using both lifting trucks and cranes have the lowest efficiency scores. One explanation could be that the terminals using cranes are handling containers to a larger extent.

Length of track

A study of working operations at the terminals showed that the length of track is of no decisive importance. None of the terminals visited had tracks long enough to receive a full train (625 metres). Short tracks involve shunting, but when the tracks are short, lifting trucks and cranes do not have to travel as far in the longitudinal direction of the tracks to handle the load units. There could be certain productive advantages of having many short tracks compared to having fewer and longer ones.

Terminal area

Terminal area is the variable that shows the largest variation between the terminals. The correlation between this variable and the total number of handling units is considered to be very high.

Table 6. Terminal area and number of handled load units (all 51 terminals)

	T9	T10	T40	S6	S7	S11
Total number of load units/terminal area	2.35	2.38	4.56	0.69	0.51	0.89
Trailer/terminal area	0.38	0.33	0.0	0.25	0.01	0.37

The German terminals definitely make better use of space than the Swedish terminals do. When the Swedish terminals were visited, it appeared that especially trailers require much space. In the same space occupied by one trailer, 3 - 4 containers could be deposited. This shows that trailers do affect efficiency and require more space. It is characteristic of the German terminals that they have assigned areas for special purposes. Truck areas and depot areas are separated, which minimises the use of land.

When comparing the partial ratio trailer/TAR with achieved DEA score for all examined terminals a negative correlation was found (-0.19). The correlation is small but the sign indicates that more trailers will lead to less inefficiency. The advantage of a trailer is that it only needs to be handled once by the handling equipment. As soon as it stands on the ground, the truck is able to couple the trailer. This advantage is not measured by using DEA.

Capacity utilisation

High use of capacity will lead to high DEA efficiency score. The data used to measure efficiency is collected on a yearly basis. Variations over time are not considered. To investigate capacity utilisation over time can give a more detailed explanation why some terminals are more efficient than others.

Table 7. Capacity utilisation per hour

	T9	T10	T40	S6	S7	S11
Average capacity (total number of load units/hour)	8.5	8.3	1.8	7.8	1.6	4.6
Max. capacity utilisation (tot. numb. of load units/peak hour)	40	35	10	40	10	35
Max. capacity utilisation/average capacity utilisation	4.7	4.2	5.6	5.1	6.3	7.6

The first line in Table 7 shows average capacity utilisation. The second line shows the average capacity utilisation during peak hours. The managers at the terminals were asked how many load units were handled during peak hours. The last line is the ratio between line 2 and line 1. This ratio gives a hint about the difference between peak hours and average hours. The terminals S7 and S11 seem to have a larger difference between the peak hours and average hours compared to the others. A lot of free capacity probably exists between peak hours. The results are not clear but a slight distortion between the terminals is discernible. The variation over time could be one explanation why some terminals make better use of their capacity. The inefficient terminals could be, to a larger extent, dimensioned for peak hours which will lead to over-capacity.

CONCLUSIONS

This study is aimed at showing how DEA can be useful as a decision support tool for combined transport terminals. The DEA method was, therefore, applied to terminals in three steps. In step 1, efficiency scores for each terminal were calculated and presented. In this step, DEA was proved to be a method suitable for revealing inefficiencies. In this step, the DEA results were studied in detail to reveal interesting factors that could influence the received efficiency scores. In step 2, only quantitative data received from the use of DEA was analysed. In step 3, examples of the adequacy of the DEA measurement were analysed. This was achieved by six case studies of different terminals in order to show the possibilities of implementing, in real situations, efficiency improvements suggested by the results from the DEA measurements.

This course of action shows how information can be created. It also takes into consideration how the terminals must act to improve their efficiency. One aspect which has not been taken into consideration is how this information should be distributed among the terminals. Since no competition, internationally or nationally, exists among the terminals, the access to information would not be a problem. Also, the information gathered does not require a great effort. The distribution of information would, therefore, be a small problem.

Another subject not discussed is how often this comparison should be made. It depends on the choice of variables. In this study, variables describing the dimension of the terminals are used. These variables are not easy to affect or change in a short perspective. The choice of variables concerns reinvestments at or new investments in terminals. If other variables are chosen the comparison could perhaps be made more on a day-to-day basis.

This study has shown that efficiency improvements at combined terminals are possible. The aim of this study was to find a course of action to bring forward incentives for efficiency improvements. Hereby, a so called artificial competitive situation can be reached. Information containing incentives for improvements of efficiency ought to be distributed among the terminals. These types of incentives do not always exist in a monopolistic situation. The purpose of distributing the information is to support the managers at the terminals both in investment and day-to-day operation decisions.

To improve the "quality" as far as the operation of terminals is concerned, goes hand-in-hand with a desire to increase efficiency. If the efficiency of the terminals can be improved, this could mean that the total costs for using combined transports will decrease. However, the terminal handling is only a part of the production of the combined transport service. Terminals, road transport and rail transport together form a combined transport system. Rendering the terminal handling more efficient without considering the surrounding system could imply a suboptimization for the whole combined transport system.

The interpretation of the efficiency measures made with DEA is rendered more difficult by the fact that the terminals are shown in their "best light" and, therefore, major inefficiencies could be concealed behind the efficiency measures received. At the same time this means that placing terminals in order of precedence loses significance to a certain degree. In the quantitative analysis of the chosen terminals, step 2, new procedures were applied. Each terminal is compared, in the first place, with its reference terminals and its peers, i.e. efficient terminals, which are included in the reference terminal. The utilisation of the different resources and their effect on efficiency were analysed. The results could be used as a guide when applying measures in order to improve efficiency.

The case studies, step 3, have shown why certain resources are being used inefficiently. At the same time they have also shown that it is possible to improve efficiency. Hereby they confirm certain results from the DEA measurements and the quantitative analysis made for each terminal. The observations made at each terminal show that considerable increases in efficiency can be achieved if the resources are used more adequately and, at the same time, this will not affect the service level of the terminals.

One advantage of the DEA method is that it can treat non-monetary data in order to obtain a measure of overall efficiency. This was the decisive reason why DEA has been used in the study of combined transport terminals where real costs are hard to define. The advantage of the DEA method lies in its ability to reveal inefficiencies but it does not explain why these inefficiencies occur. To reveal inefficiencies on an overall basis and thereby to be able to compare and rank the examined units is, here, considered to be one element in a bench marking strategy. To have more use of the DEA results in a decision situation, a procedure must follow to evaluate the causes of inefficiency.

REFERENCES

Andersson T.D. (1996). Traditional Key Ratio Analysis Versus Data Envelopment Analysis: a Comparison of Various Measurements of Productivity and Efficiency in Restaurants. N. Johns (ed.). *Productivity Management in Hospitality and Tourism*, Cassell, London.

Baker, J.A. (1989). Measures of Effectiveness for LTL Motor Carriers. *Transportation Journal*. Winter, 1989.

Banker, R.D., Conrad, R.F., & Strauss, R.P. (1986). A comparative application of data envelopment analysis and translog methods: An illustrative study of hospital production. *Management Science, 32*(1), pp. 30-44.

Banker, R.D. (1984). Estimating most productive scale size using data envelopment analysis. *European Journal of Operational Research*, 1994:35, pp 35-44.

Bergendahl, G. (1993). *Allfinanz, Bancassurance and the future of Banking*, Memorandum (School of Economics), Göteborg.

Charnes, A., Cooper, W.W. & Rhodes E. (1978). Measuring the Efficiency of Decision Making Units. *European Journal of Operational Research*. 2, 429, 1978.

EU White Paper. (1994). Transport. Workprogramme. Edition 1994. European Commission.

Farrell, M.J. (1957). The Measurement of Productive Efficiency. *The Journal of Royal Statistical Society*, 120, Part III, Series A.

Fried, H.O., Knox Lovell, C.A., & Eeckaut, V. (1992). Evaluation of the Performance of US Credit Unions. *Journal of Banking and Finance*, 17, pp. 251-256, North Holland.

Försund, F.R. (1992) A Comparison of Parametric and Non-Parametric Efficiency Measures: The Case of Norwegian Ferries. *The Journal of Productivity Analysis*, 3, pp. 25-43.

Gathon, H.J, & Perelman, S. (1992). Measuring Technical Efficiency in European Railways. A Panel Data Approach. *Journal of Productivity Analysis*. 1/2, June 1991.

Greenberg, R., & Nunamaker, T. (1987). A generalised multiple criteria model for control and evaluation of nonprofit organisations. *Financial Accountability & Management, 3*(4), pp. 331-342.

Hjalmarsson, L., Bjurek, H. & Isaksson. (1991). *Produktivitetsmätning och effektivitet inom svensk bankverksamhet*. SOU 1991: 78

Jennergren, L.P. & Obel, B. (1986). *Forskningsevaluering - eksemplificeret ved 22 ökonomiske institutter*. Økonomi og Politik. 85/86, pp.276-285.

Kumbahkar, S.C. (1989) Economic Performance of US Class 1 Railroads: a stochastic frontier approach. *Applied Economics*, 1989, 21, pp. 1433-1446.

Parsons, L.J. (1994) Benchmarking for marketing producitivity. *The First International Workshop on Service Productivity*. Brussels, Belgium, Oct. 3-4, 1994.

Prior, D., & Sola, M. (1994). Output Measurement and Productivity in Hospitals. *The First International Workshop on Service Productivity*. Brussels, Belgium, Oct 3-4, 1994:

Sjögren S. (1996) *Effektiva kombiterminaler - En tillämpning av DEA*. (Avhandling för doktorsexamen, FEK. GU). BAS. Göteborg.

Tulkens, H. (1992). *On FDH Efficiency Analysis: Some Methodological Issues and Applications to Retail Banking, Courts and Urban Transit*. Center for Operations Research and Econometrics. Université Catholique de Louvain.

12

PRINCIPLES FOR THE CHOICE BETWEEN ROUTINE MAINTENANCE AND PREMAINTENANCE IN A NETWORK OF PUBLIC ROADS

Göran Bergendahl
Henrik Edwards
Peter Svahn

ABSTRACT

This paper is devoted to the optimal choice between routine maintenance, preventive maintenance and reinvestment for a road network. The study is based upon an assumption that routine maintenance and preventive maintenance are two alternative actions to control and to reduce a deterioration of a road network until the date of reinvestment. The study will be concerned with how an agency responsible for the maintenance of a network of public roads may obtain an optimal strategy based on these two alternatives.

The paper starts in with an analysis of the relation between on one side the maintenance activities and on the other the standard or the capacity of the road. Provided this basis we formulate in a mathematical model for the optimal choice of maintenance activities up to the date of reinvestment. Then the model is utilized for numerical experiments. Practical facilities to generate data for a regular use of such a model are discussed. Then we investigate how such a model may be implemented into the regular studies of the Swedish National Road Administration (SNRA). Finally, the final conclusions concerning this total framework are given.

Keywords: Premaintenance, deterioration, optimal dates, myopic rule, pavement management systems

INTRODUCTION

In most modern countries the public roads represent an infrastructure with substantial capital assets. In the future there will be an increasing emphasis concerning the proper maintenance of this capital, mainly because of its magnitude but also because the volume of new investments will show a lower rate of growth than during the past three decades. A large number of factors will influence the performance, utilization and maintenance of the roads. Examples of such factors are user prices, user costs, rates of usage, type of usage, weather and climate as well as physical and chemical deterioration over time. All these factors will influence the *capacity* of a road. The public agency being responsible for the maintenance will then have to choose between different activities like *routine, remedial or preventive maintenance* as well as *reinvestment*. That choice must be based upon social costs and benefits as well as budget constraints, financial alternatives and minimum service requirements.

Consider a government or a municipal administration that is responsible for the investment and the operation of a road network. Such a "Road Administration Agency" must be concerned about the deterioration of the network. It must devote considerable efforts to the choice of maintenance policies as well as to the best dates for replacement.

This study will not be devoted to investments in new roads or to the expansion of existing roads in order to meet a substantial growth of traffic volume. Instead it will be concerned with the optimal operation and reinvestment in existing road facilities. It will be devoted to the optimal time-phasing of premaintenance and reinvestment activities. We will assume that the basic infrastructure facilities like roads and bridges will have an extremely long life if they are repaired and maintained on a regular basis. For example, many streets in our cities have not been replaced in decades. Instead, the main activity of the road administration has been to perform resurfacing such as blading, regravelling, bituminous surfacing, etc. This study will develop this choice of strategies in order to determine the efficient *timing* of activities that will preserve and improve the *standard* of a road. Therefore, we assume that the basic choice for such an agency over a time period is between *routine* maintenance, *preventive* maintenance and *reinvestment*. That will become a basis for the following analysis of the optimal size and timing for road maintenance. In so doing, we make use of the experiences from earlier studies like Gertsbakh (1977), Haas & Hudson (1978), Hatry & Steinthal (1984), Paterson (1985) and Moss (1985).

The importance of regular inspection and state development assessment of roads has been discussed by O'Flaherty (1988). Frequent damage types and available maintenance measures have also been covered. Many examples of the costs involved in the maintenance operations were given in e.g. Haas and Hudson (1978). A relatively short introduction to maintenance management and operations was presented by Oglesby and Hicks (1982). A general discussion of the operations management of maintenance, including how to reach reliability and maintenance objectives and how to carry out state measurements and forecasting, has been provided by e.g. Moss (1985) and Smith (1988).

There exist many alternative ways to carry out state development assessments of roads and streets. Salter (1988) presented many technical measurement alternatives, and some causal relationships between the road surface deformation as a function of the accumulated number of vehicle axle passages as well as the vehicle propensity to skid given the road surface condition and the vehicle speed.

NCHRP (1985) analyzed the decision of maintenance measure alternative in view of a life cycle cost perspective, where the state development assessments were important for suitable timing of maintenance operations. Maintenance techniques adapted to American conditions were presented in NCHRP (1981). This publication also stressed the needs of empirical data concerning relationships between costs and state levels, but also the needs of methods for deciding on the optimal maintenance execution.

Ullidtz (1987) has presented a thorough physical and mechanical analysis of the impact on the road construction and surface caused by vehicle loads, materials, climate etc. The result was used as input to an optimization model that allocates available resources (e.g. the budget) to the various road sections (objects) in the best manner possible. The choices among the objects were governed by the user costs, maintenance costs etc., where the user costs were assessed according to the World Bank report (Paterson, 1987). Empirical state development interrelations were put into a model for a consequence analysis that is utilized for estimating the maintenance costs as a function of alternative maintenance policies.

Gertsbakh (1977) discussed the implications of uncertainty and probability analysis when carrying out maintenance management. A thorough analysis of various issues related to preventive maintenance was presented by Jorgenson et al. (1967). Of particular interest is a strategy for maintenance objects with unknown error probabilities and adaptive maintenance policies.

THE CHARACTER OF ROAD MAINTENANCE

The standard of a road may be measured in many different ways. One class of the measures is seen from the point of view of the producer of the road services, that is the road administration. Examples are skid resistance, transverse evenness, characteristic deflection, and longitudinal evenness which may be combined into an overall standard index (see e.g. Lemlin et al. 1989). Another class of measures is more concerned with how good and comfortable the road service is to the users, that are the drivers and the passengers. The lower the standard the worse the road will be to the transportation of individuals and freight. The "Present Serviceability Rating" (PSR) and the "Riding Comfort Index" (RCI) are both examples of such consumer oriented measures (see e.g. Hatry & Steinthal, 1984 and Paterson, 1987).

In this paper it is stressed that one single measure (S) of the standard of a road has to be used to obtain a joint strategy of routine maintenance, premaintenance, and reinvestment. It is an important task to plan for this standard over time in order to maximize social surplus. This implies that the benefit of the producer (the road administration) has to be balanced to the benefit of the consumers (the drivers). Consequently, the measure of the standard on a road is to be used as a basis to calculate:

  a. the repetitive cost of routine maintenance,

  b. the irregular cost of preventive maintenance and reinvestment, and

  c. the cost and benefit for the consumers that use the road or street.

A general assumption is that year by year the wear and tear will reduce the standard of a road and that the amount of reduction in standard level will accelerate up to the date of the preventive maintenance or the reinvestment. That means that the standard level of a road will follow a degressive curve over time up to a date tp when an action is taken in terms of preventive maintenance, that is resurfacing, reseal or overlay (Figure 1).

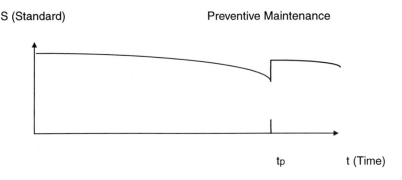

Figure 1. The development of the road standard over time

At the date tp of preventive maintenance or reinvestment there will be a sudden improvement in the standard. The user of the road services - the Road Administration - must observe a change in the outlays for road maintenance. And onwards, the higher the standard, the lower the costs for routine maintenance (patching) and the lower the future outlays for preventive maintenance and reinvestment. After an improvement of the standard has been made, vehicle operating costs as well as costs for patching will grow slowly for many years year by year up to the date when the preventive maintenance or reinvestment is completed. And for the driver the serviceability is reduced year by year in spite of using routine maintenance. Paterson (1987, p. 114) has given a good illustration of this mixture of strategies over time. Year by year the serviceability is reduced in spite of using routine maintenance. Then preventive maintenance is introduced first in terms of a resurfacing of the road and then as a rehabilitation of it (see Figure 2).

The task for the Road Administration will then be to determine the optimal combination of outlays for routine maintenance, preventive maintenance and reinvestment. Hatry & Steinthal (1984, p.6) have identified seven basic maintenance strategies in use by operating agencies. They are not mutually exclusive but may be applied in combination. They are:

1) Only do crisis maintenance;

2) Worst first;

3) Opportunistic scheduling;

4) Prespecified maintenance cycle "standards";

5) Repair components "at risk";

6) Preventive maintenance;

7) Reduce the demand for wear and tear on the facility.

The routine maintenance will cover strategies 1, 2, 3 and 4, while the preventive maintenance is identical to strategies 5 and 6. An additional eighth strategy will be:

    8) Reinvest.

With such a separation, routine maintenance is an activity that will "keep up" with the wear and tear on the facility. It would be expected that for a given traffic load (L), the lower the standard of the road (S) the larger will be the outlays for routine investment $c_r(L,S)$. And given the standard, the larger the load the higher the cost of routine maintenance. This behavior may be illustrated in terms of two cost curves - one with the cost of routine maintenance as a function of the standard (Figure 2) and the other with that cost as a function of the traffic load (Figure 3).

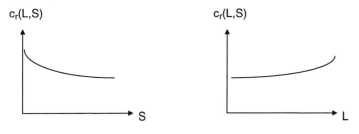

Figure 2. The cost of routine maintenance as a function of the standard (S)

Figure 3. The cost of routine maintenance as a function of the traffic load (L)

It may be assumed that even if the routine maintenance is performed in an acceptable way, the comfort of driving on the road may be reduced year by year as an effect of the tear and wear. This indicates that the social effects of the deterioration of a road will not only have to be measured in terms of higher costs of routine maintenance but also in terms of an index for comfort and convenience.

On the other hand, preventive maintenance and reinvestment will substantially improve the standard and the serviceability. One may expect that the outlays for preventive maintenance activities consist of two different parts, namely:

a) a fixed "set up" cost for starting preventive maintenance activities;

b) a cost for resealing, overlay, resurfacing and rehabilitation that is expected to be proportional to the degree of improvement in the standard.

Given this background, the choice of date t_P to initiate an action of preventive maintenance is an important task both in theory and practice. Here we take an economic point of view in order to determine the optimal time-phasing of routine and preventive maintenance. In the next section we will construct a model to be used to determine this time-phasing under different assumptions concerning cost escalation and servicability.

A MODEL FOR THE OPTIMAL CHOICE BETWEEN ROUTINE AND PREVENTIVE MAINTENANCE

The mathematical model developed in this section is devoted to the optimal time-phasing of infrastructure maintenance. In the beginning we assume that we plan for maintenance from a date $t = 0$ into infinity. Later on we will introduce a time horizon, determining when the facility has to be completely replaced.

Central for this analysis is the assumption that the *serviceability* or *standard* could be measured by one *state variable* which is denoted S. This variable may measure a Present Serviceability Rating (PSR), a Riding Comfort Index (RCI) or any other quality of the infrastructure facility.

Initially the state variable has a value of S_0. For simplicity we assume that this level is the best one that could be obtained. Year by year the state will deteriorate. Then the time development of the standard will look like the RCI- and PSR-curves of Figure 2. It will also correspond to the mirror image of the road roughness curve in Figure 3. Consequently, the state may be assumed to follow, for example, the path:

$$S_t = S_0 - S_0(e^{\alpha t} - 1) \tag{1}$$

for $t > 0$. Here S_t/S_0 is the degree of deterioration and α (>0) is a deterioration rate parameter.

The *optimal date of preventive maintenance* will be called T. We then assume that at date T actions are taken to restore the facility to its initial state S_0. After that the deterioration will follow an identical, repetitive path and we will have a new date for preventive maintenance at 2T. Without changes of the system and parameters this process will continue indefinitely (see Figure 4).

State (S_t)

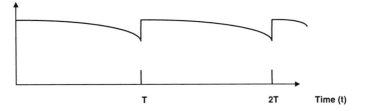

T   2T   Time (t)

Figure 4. The state of a facility as a function over time

The *cost of operating the facility* during the time period from year 0 to year T will then consist of three components. The first one is an annual fixed cost F_1. This cost concerns the basic operation of the facility and is assumed to be independent of its condition. However, the second and the third cost components are defined as the incremental costs of the deterioration of the facility. The second one is the incremental cost of routine maintenance and the third one is the incremental cost for the customer to use the facility. So both the second and the third cost components are assumed to be proportional to the standard and therefore to the state S_t. Consequently, the incremental cost of operation $c_1(t)$ at date t will be formulated as:

$$c_1(t) = \gamma(S_0 - S_t) \tag{2}$$

Inserting (1) into (2) gives:

$$c_1(t) = \gamma S_0(e^{\alpha t} - 1) \tag{3}$$

This formulation implies that the operating cost will also increase exponentially up to the date (T) of preventive maintenance (see Figure 4).

The *cost of preventive maintenance* will consist of two parts. The first one, $c_2(T)$, is assumed to be *proportional to the deterioration* $(S_0\text{-}S_T)$:

$$c_2(T) = \beta(S_0 - S_T) \tag{4}$$

Inserting (1) into (4) gives:

$$c_2(T) = \beta S_0(e^{\alpha T} - 1) \tag{5}$$

The other part of the preventive maintenance is *a set-up cost* i.e. a fixed cost associated with the organization of this maintenance activity. This set-up cost (F_2) will then be independent of the deterioration.

The purpose is now to search for the date T that will minimize the overall discounted costs (C). Over the period (0, T) this cost will become:

$$c(0,T) = \int_0^T [c_1(t) + F_1] e^{-rt} dt + [c_2(T) + F_2] e^{-rT} \tag{6}$$

Now, assume that preventive maintenance acitivies will take place in cycles of length T, that is at dates T, 2T, 3T, 4T,....In case this process continues indefinitely, one will obtain the total discounted cost as:

$$c_*(0,T) = c(0,T) + e^{-rT} c(0,T) + e^{-2rT} c(0,T) + ... = \frac{c(0,T)}{1-e^{-rT}} \tag{7}$$

The overall discounted cost is finally obtained by inserting (3) and (5) into (6) and then (6) into (7) according to:

$$c_*(0,T) = \frac{1}{1 - e^{-rT}} \{ \int_0^T [\gamma S_0(e^{\alpha t} - 1) + F_1] e^{-rT} dt + [\beta S_0(e^{\alpha T} - 1) + F_2] e^{-rT} \} \tag{8}$$

Here F_1 will be fixed year by year up to infinity. Therefore it will have no influence on the optimal date T of preventive maintenance. Thus, we will eliminate it from the analysis onwards. Dropping F_1 and minimizing C $(0, \alpha)$ with respect to T under the assumption that T>0 gives:

$$-e^{(\alpha-r)T} \alpha S_0 (\frac{\gamma}{\alpha - r} + \beta) + e^{\alpha T} S_0 (\gamma + (\alpha - r)\beta) + r (\frac{\gamma S_0}{\alpha - r} + \beta S_0 - F_2) = 0 \tag{9}$$

Approximately set:

$$e^{(\alpha - r)T} = 1 + (\alpha - r)T + \frac{(\alpha - r)^2 T^2}{2} \tag{10}$$

$$e^{\alpha T} = 1 + \alpha T + \frac{\alpha^2 T^2}{2} \tag{11}$$

Then we obtain the optimal date T* of preventive maintenance as:

$$T^* = \sqrt{\frac{2F_2}{\alpha[\gamma + \beta(\alpha - r)] S_0}} \tag{12}$$

This approximate solution displays a set of interesting properties for the optimal date T*:

    a) T* will increase with larger set-up costs, F2;

    b) T* will decrease with larger deterioration rates;

    c) T* will decrease with higher proportional factors, β,in the preventive maintenance cost expression;

    d) A necessary condition for preventive maintenance is that

$$\gamma > \beta\rho - \beta\alpha \tag{13}$$

This implies a "Myopic" rule stipulating that preventive maintenance should only be considered when the cost of operation (γ) exceeds the difference between the interest (r) and the deterioration rate (α) multiplied by the cost (β) of improvement in terms of resealing, overlay, resurfacing and rehabilitation.

The calculation above is based on the assumption of an infinite life and constant problem parameters. In a more realistic case the life of the facility will be finite and one has to introduce a date nT at which reinvestment has to be performed. The calculation of an optimal T* will then become a little more complex, but the same conditions as above have to be fulfilled.

Obviously, the task to estimate the optimal date of premaintenance will depend on three essential issues:

• the rate α of deterioration of the standard (state) S of a road;

• the cost of routine maintenance at time t as a function of the present standard St;

• the cost of premaintenance proportional to the need for upgrading S0-St.

Below we will demonstrate how to perform such an estimation in principle and in practice.

NUMERICAL EXAMPLES

Let us now introduce some numerical examples based on real life data. A development of the model regarding the operating costs in equation (2) is then made in order to obtain a more accurate model.

Jönsson (1990c) has studied how the basic data available in the computer based system "Galant" can be used in a dynamic programming context for determination of the optimal time intervals for preventive maintenance. Those data were given in costs per square meters per year (SEK per m2 & year). The alternative states of the system were identified as numbers from the set (1, 0, -1, -2), where "1" stands for a road surface in a very good condition and "-2" for one satisfying the minimum service level.

In order to exemplify the model with some numerical results, we use the input data from Jönsson (1990c). The routine maintenance cost was estimated as 0.3 SEK/m2 & year for a road in good condition (state level 1) and increased to 0.7 SEK/m2 & year for a road in the lowest acceptable condition (state level -2). The marginal cost for the road users was estimated zero at state levels 1 and 0 and it rises to 2.8 SEK/m2 & year, comprised of 1.00 for accidents, 0.50 for travel time and 1.30 in vehicle costs, when the state level goes down to -2. The total operating costs are summarized in Table 1.

Table 1. Actual operating costs

State level	1	0	-1	-2
Operating cost	0.30	0.50	1.75	3.50

For obtaining the same costs in the model when the state level goes from 1 to -2, that is a difference of $S_0 - S_t = 3$, we get the cost coefficients $F_1 = 0.30$ and $\gamma = 1.10$.

With the same argument we obtain a fixed cost, F_2, for preventive maintenance in the order of 40 - 50, and a proportional cost β in the neigbourhood of 5 - 10. Suitable values are then $F_2 = 40$ and $\beta = 10$ or $F_2 = 50$ and $\beta = 5$.

Assuming that a 20-year cycle results in a state level deterioration from $S_0 = +1$ to $S_{20} = -2$ we obtain the deterioration rate from $20\alpha = \ln 3$, i. e. $\alpha = 0.055$. Should the cycle be only half that long (10 years), then we get $\alpha = 0.11$.

Finally, we use a real rate of return of 4 %, i.e. r=0.04.

In order to study the *robustness* of the optimal date T* of premaintenance we test our model

a) for *two* different rates of deterioration ($\alpha = 0.055$ or 0.11);

b) for *two* alternative fixed costs for preventive maintenance ($F_2 = 40$ or 50);

c) for *two* alternative variable costs of preventive maintenance ($\beta = 5$ or 10);

d) for *three* alternative costs of operation ($\gamma = 0.15$, 0.60 or 1.10).

The results of these test runs are presented in Tables 2, 3, 4, and 5, where one will find:

1) the optimal date of premaintenance T* according to (12), that is T_1;

2) the optimal date T* obtained through the use of the numerical interval bisection method, that is T_2;

3) the relative difference "R" given (in %) in the value c (0,T) of the objective function in equation (8) and obtained by using the two T-values above. That means R=100 $[c_*(0,T_1) - c_*(0,T_2)] / c_*(0,T_2)$.

Table 2. Premaintenance dates for a low degree of deterioration and a high cost of premaintenance ($\alpha=0.055$ and $F_2 = 50$)

γ		.15	.60	1.10
	T₁	39.34	89.89	1.10
5.00	T₂	27.72	47.60	33.66
	R	10.93	50.92	18.64
β				
	T₁	77.85	49.24	38.14
10.00	T₂	43.70	32.48	27.09
	R	33.63	15.06	9.39

Table 3. Premaintenance dates for a low degree of deterioration and a low cost of premaintenance ($\alpha=0.055$ and $F_2 = 40$)

γ		.15	.60	1.10
	T₁	80.40	46.42	35.18
5.00	T₂	44.56	31.18	25.53
	R	38.24	14.22	8.39
β				
	T₁	69.63	44.04	34.11
10.00	T₂	40.79	30.05	<u>24.95</u>
	R	25.15	11.41	7.17

Table 4. Premaintenance dates for a high degree of deterioration and a high cost of premaintenance ($\alpha=0.110$ and $F_2 = 50$)

γ		.15	.60	1.10
	T₁	42.64	30.93	25.04
5.00	T₂	21.29	17.57	<u>15.34</u>
	R	106.95	48.25	29.89
β				
	T₁	32.70	26.44	22.47
10.00	T₂	18.19	15.90	14.27
	R	48.74	30.41	21.53

Table 5. Premaintenance dates for a high *degree of deterioration and a* low *cost of premaintenance ($\alpha=0.110$ and $F_2 = 40$)*

γ		.15	.60	1.10
β	T_1	38.14	27.67	22.40
5.00	T_2	19.95	16.37	14.24
	R	75.90	35.79	22.59
	T_1	29.25	23.65	20.10
10.00	T_2	16.97	14.77	<u>13.22</u>
	R	35.53	22.65	16.24

The results show, as expected, that the second order approximation of equations (10) and (11) give qualitatively correct results, but, in general, unrealistic quantitative results. However, the optimal T2-values (see point 2 above) are easily obtained. A disturbing fact, in comparison with reality, is that the optimal T2-values in Tables 2 and 3 are larger than 20 years which was the initial maximal period of time before preventive maintenance must be made due to the agreed service levels. The situation is the same in Tables 4 and 5, except that the maximal cycle time is 10 years. The underlined T2-values in the tables correspond to the most suitable combinations of F_2 - and β-values.

A possible explanation to the fact that the T2-values are longer than the maximum cycle time is that equation (2) in this case does not describe the operating cost development over time very well. By using a simple second-order polynomial:

$$c_1 (t) = F_1 + \gamma(S_0 - S_t)^2 \qquad (13)$$

we obtain in this example a very good fit to the data of Table 1 with $F_1 =0.239$ and $\gamma=0.364$.

The overall discounted cost, corresponding to equation (8), with $c_1 (t)$ according to equation (13) will be:

$$c_. (0,T) = \frac{1}{1 - e^{-rT}} \{ \int_0^T \{F_1 + \gamma[S_0(e^{\alpha t}-1)]^2\}e^{-rt}dt + \{\beta S_0(e^{\alpha t}-1)+F_2\}e^{-rT} \} \qquad (14)$$

Using the same input data as before, the T-values minimizing c (0,T) are obtained through a simple sequential search. The results are presented in Tables 6 and 7.

Table 6. T-values minimizing c (0,T) when the deterioration rate is $\alpha=0.055$

β\ F_2	40	50
5	24.5	<u>25.5</u>
10	<u>24.0</u>	25.5

Table 7. T-values minimizing c (0,T) when the deterioration rate is $\alpha=0.110$

β\ F_2	40	50
5	13.0	<u>14.0</u>
10	<u>12.5</u>	13.5

By comparing the underlined values of Tables 6 and 7 with the corresponding values (on F2 and β) in Tables 2 - 5 we find that the resulting optimal T-values are approximately the same. This is explained by the fact that it is the marginal costs that determines the best T-values. However, since the operating cost function from time 0 until time T is very different, the estimated costs in the two cases will differ substantially.

In order to analyze the problem of preventive maintenance in this manner we start by identifying the availability of

- Routine maintenance;
- Preventive maintenance;
- Reinvestment;
- Set-up costs and incremental cost of operation.

Next we turn to different constraints that may limit our choice of decision variable as well as to conditions describing the state (St) of the system. All these constraints are functions of the variable values, both endogeneous and exogeneous variables. In short, relevant constraints may be:

- Budget constraints;
- Maintenance capacity constraints;
- Minimum service requirements;
- Juridical restrictions;

- The state of the system and its development over time (the state can be expressed in terms that describe the condition of a facility, such as for example a road or a water pipe, in various aspects).

Within the degree of freedom given for choice of decision variable values, we want to minimize the value of the objective function. By assuming that the deterioration of a facility will induce a cost to the user, we will search for the minimum of the discounted social cost.

In practice it is usually very difficult to identify all the relevant variables and their impact on the different relationships expressed in the constraints and in the objective function. An alternative approach will then be to express the social costs in indirect terms. An example is to introduce a minimal desired service level for the service, instead of expressing it in terms of social benefit (in e.g. monetary terms).

DATA ANALYSIS

We consider it to be of vital importance to describe the *state* of the system and its development in a relevant manner. For example, the state for a section of a road could be defined in terms of rutting levels, number of cracks per meter or as a bearing capacity. In a water distribution system the state variables for the various water pipes could reflect the failure rate and the ratio of the current diameter to the diameter of a new pipe (a clogged up rate).

There are a large number of factors affecting the state of the system, of which some of the most important are:

- usage rates;
- type of usage;
- maintenance activities;
- climate;
- physical and chemical deterioration over time.

Given a dynamic description of the state of the system, we may be able to relate an important part of the social costs to the state of the facility. The user costs may be described in terms of travelling time and vehicle usage. Both types of costs increase when the state of the facility deteriorates. Costs for routine maintenance can also be presented as a function of the state.

Regarding service requirements, they may be expressed as minimum (or maximum) levels allowed for different state variables. As an example consider a road section where the maximum allowed rutting level is set to 20 mm.

Hence, given an appropriate state description of the system and its dynamic behavior, many important aspects of the overall problem can be reflected. Other factors in the problem such as user prices and financing alternatives have a direct impact on the function of the infrastructure facility. This follows from the fact that the prices affect the system utilization. In a corresponding manner the amount of maintenance carried out depends on financial constraints. If they are substantial, preventive maintenance activities may have to be postponed.

The model is based upon the experience made during a large study of the maintenance policies for Swedish roads and water pipes. An introductory study was carried out by Bergendahl et al. (1987). Many of the basic ideas and research topics regarding pricing and financing of maintenance activities were also outlined in that study.

A literature overview concerning theory and methods of investment and maintenance is presented in Löfsten (1989a). This was followed by an empirical study, Löfsten (1989b), of the operations of road and street maintenance and of the water distribution and sewage system in Gothenburg (the second largest city in Sweden with approximately 500,000 inhabitants).

Jönsson (1990a) has made a deep analysis of the relations between use and maintenance costs and system state development in general for the roads in an urban area. He has based that analysis upon a set of available documents (see e.g. NCHRP 1981 & 1985; Oglesby & Hicks 1982; Ullidtz 1987; O'Flaherty, 1988 & Smith 1988). Then an illustration of the allocation of maintenance to a selected section of a road is given through some numerical examples. In these examples it is assumed that the state of the system can be described in terms of a road surface condition and a bearing capacity value, that also are dependent of each other. All variable social costs are expressed as functions of the actual state plus the costs for the different maintenance actions. The maintenance alternatives are a remedial maintenance or different preventive maintenance and reinvestment actions. A straightforward dynamic programming formulation, including the use of discrete state variables, is used for solving the problem and for a generation of a set of solutions that are near-optimal over a 60 year planning horizon. Each solution indicates what maintenance actions to perform and in which periods. Furthermore, a model is suggested for how to choose a maintenance plan among a given set for each road section considered subject to the budgetary and financial constraints. In this model a number of road sections in need of maintenance are considered simultaneously. An illustration of the system state development is given in Figure 5.

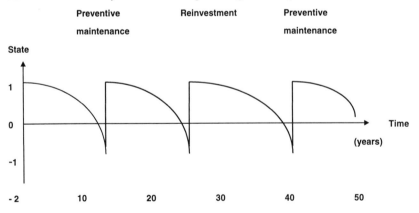

Figure 5. System state development for a road section

Jönsson (1990c) has made an analysis of a computer assisted planning system called GALANT that has been used by some local governments in Sweden. In that system a number of standard levels of service, +1, 0, -1 and -2, are defined for each section of a road, and a maintenance cycle for maintaining each standard level is defined for each road depending on the type of road and the usage rate. The social costs considered in the problem formulation are costs for marginal deviations from the basic case, and they include costs for travelling time, vehicle costs, accident costs and maintenance costs. The social costs, excluding maintenance costs, were based on conservative estimates (i. e. on the low side) of costs obtained from various empirical research results from, for example, the Transport Research Board and the Swedish Road and Traffic Research Institute. In the paper it is suggested how a maintenance cycle can be extended, at an appropriate cost, to cover more than one standard level during a full maintenance cycle that starts with raising the maintenance standard to its initial highest level and ends at a possible lower lewel. As an example the maintenance cycle could start with the standard level 0 and end at standard level -1 (before it turns into standard level -2). An example of this is given in Figure 6.

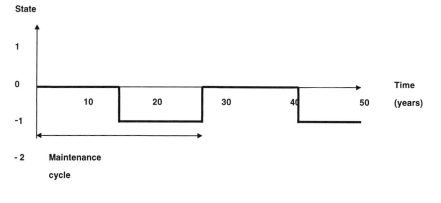

Figure 6. Extended maintenance cycle covering two standard levels

This approach has been formulated in terms of a dynamic programming model with a long planning horizon (250 years) in order to avoid that the choice of a horizon should have an effect on the optimal policy. The state variables were given in terms of an initial standard level of the maintenance cycle and its age. When the actual standard level goes down it may be necessary to improve the structure of the road in order to regain its initial level. The results when applying the model to data concerning a city in Sweden, Linköping, with approximately 120,000 inhabitants, were the lowest cost maintenance cycles possible for roads and streets that were divided into 9 classes.

The uncertainty calls for an analysis based on the inherent stochastic properties of the problem. In the overview paper by Jönsson (1989a) some stochastic optimization models are presented briefly. The traditional stochastic dynamic programming approach, for example, was used in the paper by Jönsson (1990d) discussed above. Also the recently presented scenario aggregation approach for solving multistage stochastic optimization problems is of interest.

PRACTICAL APPLICATIONS TO THE SWEDISH ROAD NETWORK

The outstanding aim of the Swedish road transport policy is to offer the citizens and the business community a satisfactory, safe and environmentally acceptable road transport system (RTS), to the minimum cost for the society as a whole. This aim will also stress that the consumer (of the road transport services) should be put in focus and offered a high degree of freedom regarding how to use the RTS.

Since an increasing part of the total cost for the RTS in Sweden is due to operation and maintenance, it is becoming more and more important to perform these tasks in an efficient manner. The budget for operation and maintenance during 1996 is 5 867 million Swedish crowns (MSEK) compared to the investment budget that is 5 000 MSEK for the same period. Because of the large amount of money spent on operation and maintenance, it is imperative to perform this task efficiently with the purpose to maximize the social benefits minus the social costs. It is also necessary that the investment plan includes the future maintenance cost during the lifetime of the investment in order to minimize the total costs of the road facilities.

To be able to plan the investment and maintenance strategies efficiently within the SNRA there is a need to know:

- The initial standard of the road transport system;
- The deterioration rate for the complete road network as well as for each single link of the road network;
- The cost to maintain a certain standard in the road network or single road link;
- The cost to upgrade the standard of a single road link.

At each date the desired standard of the network should be determined so that the consumers' marginal willingness to pay (the marginal revenue) is equal to the marginal social cost. Thus it is a prerequisite that the present standard of the roads can be evaluated and that the desired standard can be decided based upon the marginal cost and revenues for the society as a whole. Within such an analysis, the road administrator (SNRA) must develop a system approach so that the desired standard of the road transport system can be reached while minimizing the costs.

In an analysis concerning the efficiency in planning investment and maintenance, one central question concerns the over all planning of the complete road network. The task is to find where in the road network a marginal expenditure is most beneficial (to the society) and should it be spent on maintenance or investment? This task is performed in two phases. In the first phase the national road budget is divided up between the regions. Then in the second phase each region has to determine an optimal balance between investment and maintenance. These two phases should be repeated until the strategy is not only optimal for that region but also for the whole country.

The above decision-making will be made in the presence of a set of constraints that prohibits an overall unconstrained optimal investment and maintenance strategy. These constraints include information and knowledge as well as budget limits. The information/knowledge restrictions deal with the fact that there may be a lack of information and knowledge that makes it impossible to attain an optimum investment and maintenance strategy. The budget constraints deal with the fact that there may be a lack of funds that prohibits an achievement of the best investment and maintenance strategy even if there were no restrictions concerning information and knowledge.

In a strict economic perspective, a marginal expenditure for investment or maintenance should be spent on the road link that gives the highest marginal surplus for the society. This means that the SNRA should obtain funds in coherence with their marginal surplus in competition with other transport systems and other public services in the society. To avoid a complicated analysis of the SNRA's efficiency compared to other public utilities in the society, this study will only cover SNRA's efficiency under the budget restriction decided by the Swedish government. Therefore an assumption must be that the SNRA will minimize its costs for administrating the road network subject to the aims that the government has decided are to be met.

The assumption in this study is that if the SNRA has to meet the minimum standards for different road categories set up by the government, the social surplus is maximized. This implies that the SNRA is supposed to choose between strategies of investment, premaintenance and routine maintenance in order to maximize social surplus. Under these assumptions the road standard will depend on how well the SNRA uses the available funds to invest in and maintain the present road transport system.

The major aim of this application is to analyze the information and the information systems that are used within the SNRA today and to what extent it can be used to perform the maintenance task more efficiently. This aim will be fulfilled in two parts:

- In the first part the aim is to describe the organization and the division of responsibility within the SNRA;

- In the second part the aim is to describe and analyze the existing information and decision support systems and to investigate if they can be used to perform the maintenance task more efficiently.

The Swedish National Road Administration (SNRA)

The SNRA has been assigned by the government:

- To follow up the developments within the total road transport sector;

- To maintain and to develop the public road transport system;

- To administrate the car license and vehicle registration system.

In general, we focus on the second one of these three assignments. In particular we focus on the three alternatives in order to maintain the public road transport network.

During 1994 the SNRA was responsible for 97 930 km of public roads. On these roads approximately 70 % of the total traffic work was being executed (measured in vehicle kilometers, that is the number of vehicles times the number of kilometers).

To meet the assignments above, the SNRA has formed an organization that looks as Figure 7.

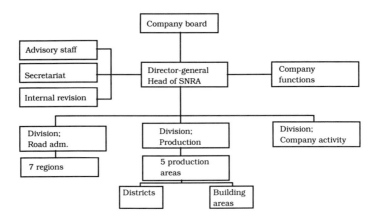

Figure 7. The organization of the Swedish National Road Administration (Source: internal material SNRA 1995)

During the 1980's the SNRA went through a major organizational change. In this change the organization was rationalized in such a way that larger geographical work units were established and the work force was reduced. Today the SNRA is divided into seven districts (earlier 28 areas) and the production division is divided into five areas. Since 1994 all the investments are subject to external competition. This means that the production division can not be certain to perform all projects available. In the same way the maintenance work is subject to external competition. However, for these activities the change has not been performed as quickly. In 1995 approximately 30 % of the maintenance task were executed under external competition between SNRA's own production division and external contractors.

The operative part within the SNRA is divided into three divisions (see Figure 7). Assignments for these divisions can be described as follows:

- The Road Administration Division. The main task consists of the planning and projecting of new roads and bridges and the improvement of existing objects. After accomplished planning and projecting the work is ordered from the Production Division or from external contractors, depending on who submits the lowest offer to relevant quality requirements.

- The Production Division is divided into five geographical production areas. These production areas are then divided into 22 production districts and 15 building areas. The task for the division includes projecting, building, rebuilding and supplementary services. These tasks are initiated and financed by the Road Administration Division. The purchase of services is done in competition with other contractors and suppliers. This means that the production division is able to offer its services outside as well as inside the SNRA.

The SNRA headquarter, situated in Borlänge, has an "economic governing mission" and (from a operative point of view) a "supportive function". The economic and governing function is formed based on the fact that funds from the government are directed to the headquarters and subsequently divided between the seven regions according to their investment and maintenance needs. The supportive function involves the task that the headquarter has to administer and to operate a number of central functions. One of these functions deals with the collection of data that is to give a major support to the investment and maintenance planning system.

Information and decision support system

The Swedish National Road Administration focuses on three different kinds of information and decision support systems, namely:

- Laser Road Surface Tester (LRST);

- Planning of Maintenance Program (PMP);

- Pavement Management System (PMS).

The most important tool for the collection of data on the standard of road links is a "Laser Road Surface Tester" (LRST). The measurement of standards is executed by a vehicle on which lasers are attached to the front and they will survey the road as the vehicle is being conveyed. The lasers will send out a beam every ten centimeters but the collected data are stored as mean values for every twenty meters. The parameters measured are track depth, irregularities (IRI), irregularities with different wavelength, crosswise inclination, horizontal curvature, vertical curvature and texture. It is primarily track depth and irregularities (International Roughness Index, IRI), that are used within SNRA when identifying maintenance objects within the road transport system. This technique has been used since 1987 and collected data has been stored for approximately 450 000 km of roads. Therefore recurrent measurements for a great number of road links exist. These periodically recurrent measurements together with the more detailed measurements done by the Swedish Road and Transport Research Institute (VTI) are potentially important data for the studies of and forecasts on the rate of deterioration for different road categories.

The SNRA is presently working with an information system for the planning of the maintenance work (PMP). This system does not contain applications for selecting appropriate rectifying actions, i.e. computations that will minimize the total cost for different actions, or information that points out the optimal action. These applications are normally available in a traditional Pavement Management System (PMS). The reason why the SNRA does not yet work with these applications is not due to lack of information but because of the fact that the system has not been developed at the same rate as the information has accumulated in the organization. The PMP is mainly built up around three databases. One consists of the information on the standard of the road transport system (mean values for twenty-meter stretches) and these values are connected to a geographical reference system. A second data base contains information about road links where each link has the same paving, year of repair and traffic intensity. These homogeneous links are approximately 700 meters. The information contains parameters like track depth, irregularities (IRI) and statistically calculated values for the deterioration rate. A third data base includes information about the planned paving i.e. the actions that will be taken to upgrade the surface in the years to come. The PMP can be categorized as a geographical information system that gives a preliminary indication on where, in the seven regions, the maintenance actions are to be taken. The users within the SNRA are closely involved in the developing of the PMP system. These developments are planned to result in a system where the possibilities for analysis are improved and as a result of this, the maintenance work will be performed more efficiently. The major deficiency in the PMP system of today is the lack of cost information concerning different maintenance and investment objects. This makes it impossible to control the life cycle costs for the maintenance and investments on a specific road link as well as to compare different maintenance strategies. Consequently the SNRA is unable to systematically control the actions that would minimize the costs over time for different maintenance categories.

The SNRA is developing the PMP system into a pavement management system (PMS) to address the problem of maintenance control. The aim is to control the maintenance activities more efficiently throughout the country.

The PMS is intended to be used as a systematic planning tool, for maintenance and reinforcement activities, that answers the questions: when, where and how this is to be done. Within the aim of the PMS lies the thought that it will be possible to predict the deterioration rate and the costs for maintenance actions at different dates. This aspect will not only cover the cost for maintaining and developing the road transport system but also the costs that are incurred by the road-user or third parties (see Figure 8).

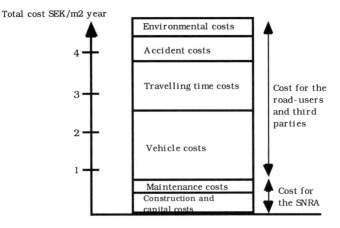

Figure 8. Different cost components for road transportation (Source: Internal material SNRA 1995)

An important function in a PMS is the fact that it should contain an optimizing model which will intend to minimize the total cost for the road transport system. Such an optimization is based upon the fact that the SNRA should try to find the standard of the road transport system which will minimize the sum of costs for the road-users and the costs for the SNRA (see Figure 9).

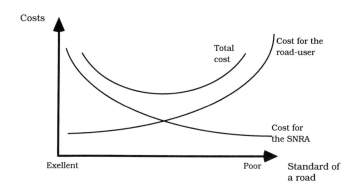

Figure 9. The principle for balancing the cost to the road-user and the cost to the road administrator (Source: Internal material SNRA 1995)

Furthermore, the intention is that the PMS will work as an aid in the following areas:

- In the collection of information that is to be used when the SNRA applies for funds from the government;
- In the process when the funds are to be divided between the seven regions;
- In the planning of maintenance actions within every region;
- In the collection of information that is to be used when the SNRA is to evaluate and choose between external contracts;
- In the process where the SNRA follows up the result from different maintenance strategies and develops the models used to minimize the total cost.

Maintenance optimization in practice

An overriding task of the Swedish road transport policy is to offer the citizens and the business community such a road transport system as they are willing to pay for. One of the aims is economic efficiency, which in this case means that the activities should be performed in a way that will maximize the surplus of the benefits to the road-users as well as minimizing the cost to the producer of the services (SNRA).

The SNRA needs regular information about the present standard of the road transport system, the deterioration rate of the complete road network as well as the single road links and the costs to keep the network/links at a certain standard (the routine maintenance), and the costs to improve the standard (the premaintenance). Given this information the SNRA should be able to raise the efficiency in the investment and maintenance planning.

There exists a set of restrictions that might make it impossible to reach an unconstrained optimal investment and maintenance strategy. The limits on information and knowledge deal with the fact that it will be impossible to measure, meter by meter, the standard of a road as well as to forecast the deterioration rate of this standard. Consequently it will be impossible to identify an optimum investment and maintenance strategy.

If the SNRA is to reduce the constraints related to information and knowledge it must obtain the information that is relevant for the models used. The SNRA must use a selection strategy that produces the decision support that makes it possible to satisfy the aims of the organization as well as the government objectives subject to the given budget restrictions.

CONCLUSIONS

Public roads represent a substantial capital asset in most modern countries. Consequently, an increasing emphasis is devoted to procedures that aim at a cost effective maintenance of that asset. This paper deals with the questions concerning the choice of strategies in order to determine the efficient timing of activities that will preserve and improve the standard of a road. The choices for a Road Administration are primarily routine maintenance, preventive maintenance and reinvestment. A presumption in this paper is that such a Road Administration has to develop principles for how to measure the standard of a road. Then it has to use this measure to obtain a joint strategy for routine maintenance, premaintenance and reinvestment.

In this paper we have developed a model for the optimal size and time for premaintenance. Given this model and given certain assumptions of costs and deterioration, it is shown that preventive maintenance should only be considered when the cost of operation (γ) exceeds the difference between the interest rate (r) and the deterioration rate (α) multiplied by the cost of improvement (β) given in terms of resealing, overlay, resurfacing and rehabilitation. Furthermore, it is obvious that the task to estimate the optimal date of premaintenance will depend on three essential issues:

-   the rate of deterioration (α) of the standard (S) of a road;
-   the cost of routine maintenance at time t as a function of the present standard S_t;
-   the cost of premaintenance being proportional to the need for upgrading S_0-S_t.

From the section in the paper dealing with the SNRA it follows that the major deficiency in the system in use today is the lack of cost information concerning different maintenance and investment objects. This shortcoming of the information system precludes the possibility to estimate the life cycle cost for the maintenance and investments for a specific road link as well as to compare different maintenance strategies. Consequently the SNRA is at presently unable to systematically evaluate the actions that would minimize the cost over time for different maintenance strategies.

In the PMS presently under development within the SNRA there is a need for recurrent information about the standard of each link in the road transport system, the deterioration rate of the complete road network as well as the single road link, the costs to keep the links of the network at a certain standard, and the costs to improve the standard. If this information were available, the SNRA should be able to increase the efficiency in the investment and maintenance planning. Even if great difficulties exist in predicting future costs for the maintenance and investments in a road transport system, it is of utmost importance to regularly collect relevant information about the standard of each link of the road network. It will not be possible to obtain an efficient maintenance strategy until this is done properly according to the principles developed in this paper.

REFERENCES

Bergendahl G, Bergendahl P A och Segelod E: Planering, prissättning och finansiering av kommunal infrastruktur ("Planning, Pricing and Financing the Maintenance of Local Government Infrastructures"), The Swedish Council for Building Research, Report R113:1987, Stockholm, 1987, (82 pages)

Gertsbakh I B: Models of Preventive Maintenance, North Holland, Amsterdam, 1977.

Haas R and Hudson R: Pavement Management Systems, McGraw-Hill, New York, 1978.

Hatry H.P. & Steinthal B.C: Guide to Selecting Maintenance Strategies for Capital Facilities, The Urban Institute Press, Washington D.C., 1984.

Jorgensen D W, McCall JJ and Radner R: Optimal Replacement Policy, North Holland, Amsterdam, 1967.

Jönsson H: Stokastiska optimeringsproblem och -metoder: En översikt ("Stochastic Optimization Problems and Methods: An Overview"), Department of Business Administration, Gothenburg School of Economics, September 1989a, (27 pages)

Jönsson H: Några synpunkter på det framtida förnyelsebehovet av VA-ledningar ("Some Views on the Future Demand for Replacement of Water Distribution Pipes"), Department of Business Administration, Gothenburg School of Economics, December 1989b, (13 pages)

Jönsson H: Diskussion av modeller för underhållsdimensionering, med exemplifiering avseende gator och vägar ("Discussion of Models for Resource Allocation of Maintenance, with Examples from Road Maintenance"), Department of Business Administration, Gothenburg School of Economics, February 1990a, (58 pages)

Jönsson H: Förnyelse av en dålig VA-ledning via omläggning eller renovering, ("Renewal of a Bad Water Pipe through Replacement or Renovation"), Department of Business Administration, Gothenburg School of Economics, March 1990b, (24 pages)

Jönsson H: Dimensionering av vägbeläggningsunderhåll baserat på Galant-systemet ("Dimensioning of Road Surface Maintenance Based on the Computer Based GALANT System"), Department of Business Administration, Gothenburg School of Economics, May 1990c, (26 pages)

Jönsson H: Analys av ersättningsinvesteringar avseende VA-ledningar med hänsyn till osäkerhet ("Analysis of Replacement Investments in Water Pipes Subject to Uncertainty"), Department of Business Administration, Gothenburg School of Economics, June 1990d, (32 pages)

Lemlin, M., Ghilain, E., Heleven, L. & Janssens, R.: Basic Principles of a Maintenance Planning System for the Belgian Road Network, Ministere des Travaux Publics, Brussels, 1989

Löfsten H: En studie av investerings- och underhållsmodeller ("A Study of Models for Investment and Maintenance"), Report 1989:288, Department of Business Administration, Gothenburg School of Economics, 1989a, (36 pages)

Löfsten H: Studier av investerings- och underhållsplanering i svensk infrastruktur med några exempel från Göteborg ("Studies of Investment and Maintenance Planning in Swedish Infrastructure with Case Studies from Gothenburg"), Report 1989:292, Department of Business Administration, Gothenburg School of Economics, 1989b, (77 pages)

Löfsten H: Principer för planering, prissättning och finansiering av kommunal infrastruktur ("Principles for Planning, Pricing and Financing of Local Government Infrastructure"), Department of Business Administration, Gothenburg School of Economics, May 1990, (72 pages)

Moss M: Designing for Minimal Maintenance Expense, Marcel Dekker, New York, 1985.

NCHRP - National Cooperative Highway Research Program Synthesis of Highway Practice: Evaluation of Pavement Management Strategies, No 77, Transportation Research Board, National Research Council, September 1981.

NCHRP - National Cooperative Highway Research Program Synthesis of Highway Practice: Life Cycle Cost Analysis of Pavements, No 122, Transportation Research Board, National Research Council, December 1985.

O'Flaherty C A: Highways Volume 2 (3rd edition): Highway Engineering, Edward Arnold, London, 1988.

Oglesby C and Hicks G: Highway Engineering (Fourth edition), John Wiley, New York, 1982.

Paterson, W.D.O.: Road Deterioration and Maintenance Effects. Models for Planning and Management, The World Bank, 1987.

Smith D: Reliability and Maintainability in Perspective, MacMillan Education, London, 1988.

Swedish National Road Administration (SNRA), Annual Report, Borlänge, 1994.

Turvey R: Economic Analysis and Public Enterprise, George Allen & Unwin Ltd, London, 1971.

Ullidtz P: Pavement Analysis, Elsevier, Amsterdam, 1987.

13

INFRASTRUCTURE INTEGRATION AND URBAN RESTRUCTURING: A CASE FOR PLANNING WITH A GEOGRAPHIC INFORMATION SYSTEM

Åke Forsström

ABSTRACT

Travel and exchange time-savings due to increased speed and co-ordinated routing and scheduling will enhance the integrated use of transport systems in the course of a few years. The systematic use of coupled time-tables in computer databases connected to a Geographic Information System (GIS) will make a multiple planning instrument. This device can function in the case of a single traveller finding his route or a location-seeking firm as well as that of a local government planning its territory. The GIS will produce maps showing the accessible spots in space determined by given restrictions of travelling and visiting time. It will thus visualize the unvisible, but nonetheless real, new transport landscape being made possible by the integrated use of fast means of transport on road, rail and in the air.

Keywords: Spatial theory, tube quality, time geography, GIS

INFRASTRUCTURE UPGRADING AND INTEGRATION

The introduction of high-speed-trains (HST), designed for special tracks, in Japan in 1964 and in France in 1980 is probably the most spectacular sign of a technological leap in the transport world of today. Air and road transport changes have been of the additional kind and development of capacity and accessibility has grown to a later stage on the innovation curve. The impressive and continuing advancement in information and communication technology (ICT) is surely not dependent on one single achievement. It is characterized by a combination of technological progresses in many contiguous telecom applications, in which capacity and quality increases are being enhanced by an ongoing, fast integration.

The very successful integration of ICT innovations leads us to ask what the likely consequences of a corresponding integration of the sea, road, railroad and air transport systems and the ICT system may be. These four transport and communication systems have some qualities in common. They are all basically old, but are being modernized in some parts of the world. Historically land and sea communications have influenced regional economic prosperity (Braudel 1989). Along with this the urbanization process got started.

A spatial expansion began, first very slowly, later at an accelerating pace. In a local context, city formation and urban land use were influenced first by railway patterns and later by the construction of a dense road network. The urban spatial morphology, originally circular, changed to the star-shape and later took on an amorphous finger-like form (Hägerstrand 1972, Giannopoulos 1992). On the intra-city level, this process is labelled urban explosion. From a regional point of view, places first developed their hinterland, later they were connected linearly. In the third stage clusters of places were joined in regional networks. Later, air transport to some degree attached the central nodes of many clusters. As the time of travelling between them is reduced, they move closer in relative terms. On the inter-city level an urban implosion has been and is taking place (see Haggett 1972).

It is thus a relevant geographical problem to ask for the spatial implications of an increased infrastructure integration in a long-range perspective. To-day, air deregulation, competition and decreased travelling cost are parts of an expected process as well as a speed and capacity leap realized by HST investment. The use of air and rail transport is believed to be stimulated by more efficient information by ICT and by time and terminal adjusted feeder transports mainly on roads. Arguments will be put forward to show that these developments will lead to accessibility changes, new patterns of contacts and increased travel and in the end a different settlement structure.

THE ACTUAL PROBLEM

It is an accepted hypothesis that infrastructure investment and innovation in separate systems will lead to changed land use. It is not the purpose of this paper to test this hypothesis further, but it will be used as one of the starting points of this discussion. The technological development, the history of geographic changes and those infrastructure extensions which up to now have been carried out seem to justify this theory.

The problem at hand concerns the spatial implications of simultaneous changes of the transport systems and information technology. More than twenty years ago IT, as being a nodeless network, was predicted to pave the way for a decentralized society (Harkness 1974). One of the means to achieve this status should be the exchange of travelling for telecommunications (Henneman and Krzyczkowski 1974). Furthermore, IT-innovations were believed to spread more quickly in rural areas. Irrespective of whether these ideas are substantial or not, the outcome of a more or less coincident extension and upgrading of the nodal networks of the sea, road, railroad and air transport systems is relevant. A nodeless telenet with a fast growing number of users has to compete with four nodal transport networks with a somewhat stagnating utility. What is the outcome in terms of centralizing and decentralizing forces regarding urban structure?

Earlier, achievements in a single transport system such as air transport have been noticed as concentrating the urban population in a time dimension, the urban implosion phenomenon. It put the light on the question of how the geographic landscape actually was comprehended by moving people. It was agreed to be a mental construction. Hägerstrand (1957) made one of the first attempts to make a mental map. He used an azimuthal logarithmic projection centred on the point of origin of a journey. Other methods of showing the influence of travel time on a map are isochrones (Clawson 1972), isochrones with areal distortion (Natkiel 1970, Wegener and Spiekerman 1995), distance transformations (Matthiessen 1993). A second problem is thus how the neighbouring activity sites of a place could be mapped with realism as a primary goal.

A hypothetical chain of consecutive events consists of infrastructure improvements, accessibility changes, a rising number of mutual tele- and face-to-face contacts, travel growth, more business meetings, growth of trade, increased inter-firm cooperation, investment decisions, and in the end a different settlement structure. I have chosen to picture a time-spatial frame within which journeys take place. I concentrate on how the accessibility changes are distributed geographically in terms of decreased travel times between cities and places. These changes can be apprehended as an emerging new transport landscape, which shall be viewed at different levels of the societal organization.

First, there is a subject problem of the extension of accessibility due to infrastructure integration. Second, there is also a method problem of constructing images of transport landscapes, which may initiate new mental maps of city landscapes.

OBJECTS, QUESTIONS AND HYPOTHESES

The objects of this paper are twofold. The first is to make probable that the integration of communicaton and transport infrastructure will create substantial accessibility increases. Questions will be answered on spatial accessibility change implications for the economically active population and the business population in the area where they operate.

The second object is to design an outline of a tool based on a database of timetables and a geographic information system and make it possible to calculate shortest time-routes between pairs of cities in urban structures and to interpret them in maps. In spite of the inadequate existing techniques it is possible to describe a non-contiguous structure in maps.

A CUT THROUGH THE THEORY OF LOCATION AND MOVEMENTS

The locations of every organized group of people in society is a matter of managing resources, products, services and movements. The technical potentials and the costs of movements have decreased substantially over time. The processes of production and consumption imply an increasing degree of transports of partial results from one processing spot to another. Also the processes of preparing value added to products and streamlining the logistics engage large numbers of professionals more and more in business meetings at distant places as the internationalization of the economy grows. The sum of all movements grows as the production itself grows, as trade, transport of goods; human mobility as migration, working, shopping and servicing journeys, social trips and tourism. If we put the environmental issues to the side, the two things that matter when considering movements are speed and time.

Time geography gives us the restrictions of man's activities in a day and night perspective. The need to recover from exhausting external activities by sleeping, eating and socializing puts a time limit to the available movement time. Of the twenty-four hours, roughly eight to ten hours remain in the long run for external activities. In order to combine the resources distributed in space in the most efficient way, individuals, the organized group, or the firm has to balance between journeys and productive/consumptive activities during these eight to ten hours. This balance has different outcomes due to the purpose of the productive activities.

The routine working engagements in and around a central place create a daily urban region (DUR). It is a stable spatial creation of activities built upon a transport system and a labour market structure. The average commuting time for a single journey in city-regions is observed to be approximately half an hour with a rather narrow standard deviation (Living Conditions SOS, 1977).

The weekly business trip is an analogue to the commuting journey from a systems point of view. As appearing once or twice a week, it can be extended to up to twelve hours of combined travel and meeting activities. The single trip door-to-door travel time might be up to four hours.

These observations have spatial implications beyond the since long observed economic regions of different kinds. The walls surrounding the urban area of the ancient city were located not to exceed a walking time of twenty minutes. Also the borders of the daily urban region were positioned more or less by the same time distance. In both cases, the street or the road network practically covered the whole regional area. On the macro level technical improvements of transport systems put important cities together, the so called urban implosion (see for example Janelle 1969 and Bonnafus 1987).

COMMUNICATION AND TRANSPORT SYSTEM INTEGRATION

Starting points for the discussion are the concepts of integration and accessibility. The journey parts of the production and business activity systems are now being rationalized by a many-sided integration. This process will be examined in some detail regarding airway, railroad, highway and telecommunication systems.

Increased air deregulation, competition and decreased travelling cost - high speed, increased carrying capacity realized by HST as well as high speed ships (HSS) - may raise and diversify accessibility of each system. Transportation is stimulated by more efficient information using telecommunications and by feeder road transports carried out on a high quality and dense road network. In which manner may integrative measures, in all these five systems, change accessibility?

The term communication and transport integration (CTI) is needed to cover these measures. It stands for a conception which needs an analysis. Integration means "act of making a whole out of parts" and also "the process by which individuals of a lower order develop into individuals of a higher order" (Collins New English Dictionary 1965, p. 524). Thus, in the first place CTI covers all efforts to diminish the sacrifices of making a journey. It is agreed both by economists and geographers that time reduction is a strategic variable. In the second place it also embraces those activities which aim at expanding the journey potentials such as capacity expansion, increased access in time and space and openness for use exemplified by added transport quality, deregulation and removal of prohibitive costs.

Accessibility is a concept which is based on the possibility to take part in human activities in space. It comprises both the proximity to activity sites and the ability to move between these sites. The concept thus enhances both the activity spectrum and the mobility (Hanson 1986). In this paper the public ability to move is studied and the varying personal mobilities are disregarded. I have chosen to use time as an indicator to which extent accessibility changes are induced by integration. In this context the journey concept is vital. A journey is made up of trips performed by different means of travel and linked by exchange passages. Though it is assumed that carrying capacity is increased and that travel cost is lowered and/or travel quality is increased, nothing is taken for granted regarding societal costs of these efforts.

From a systems approach, all communication and transport systems can be described as consisting of the elements in Table 1. The aim is to show the integrated hierarchic structure in each one of these systems with a physical network at the bottom and overlayed by a traffic system with transport and communication systems at the top. These systems are organised to serve an activity system, an organisation including a limited number of persons. The public systems on the other hand do not limit their services to the people of one organisation but to the whole population in a region.

Table 1. *Components and relations of communication and transport systems*

NETWORK	COMPONENTS	RELATIONS	SYSTEMS
PUBLIC SYSTEMS			
Physical network	Links (roads/cables) Nodes (traffic centre/ amplifier)	Physical connection	Network system
Traffic network	Network system Nodes (stopping-places, exchanges)	Supervision Signaling Payment	Traffic system
ORGANISATION BASED SYSTEMS			
Transport network	Traffic system Nodes (transfer)	Organisational ties	Transport system
Communi- cation network	Traffic system Nodes (org.-units)	Organisational ties	Communication system

Source: Forsström 1995

In the development of the infrastructure system there are short run and long run problems. To include logistic steps in the production process has been a popular economic theme for a long time. These efforts have special premises in an infrastructure system. First, the size of the fixed cost is usually very large. Second, the input units in the production process of each system are commonly in the form of services from a preceding or a lateral system. At the top of each infrastructure system consumer transport or communication services are supplied.

It is a well-known property of heavy infrastructure network constructions that they work with over-capacity, primarily regarding the basic physical network services, but also the end services tend to be supplied at a rate that is greater than demand. Capacity at the network level more or less must have an excess supply due to spatial and temporal variations of demand. Short time transport variations are coupled to the activities due to the industrial, business and household organisation of society. In the long run, the slowly spreading of household settlement followed by industry redistribution, relocates demand away from the spatial cover of the physical network. The demand basis of the system is weakened. This over-capacity problem is dealt with by differential prices, at least in theory. Problems of welfare distribution, justice arguments and growing public subsidies prevent the upgrading and modernisation of these infrastructure systems (Pucher 1995a, 1995b).

The services offered at separate levels in the system are linked to one another from bottom to top, for example from basic road service to the end-service of a specific time and route determined personal transport. The production size of these services is as a rule fixed. The supply of physical services by the network or traffic system can only to a degree be utilised as inputs in the end service production by the transport or communication system. For temporal and spatial reasons there are generally over-capacity problems also in the short run. These short-comings are not the primary problems in this paper, but they are part of the background if the circumstances are changed from administration to the constructing of a system, from economizing to investing.

With advances in communication and transport technology together with a general increase in communication and transport demand, the mentioned circumstances have changed, at least in Europe (Rathery 1993). If this situation is met by new technology-based modernization phases realized by huge investments in the four old communication and transport systems, then there is a different problem at hand, sometimes called supply-induced growth. Innovation theory says that new transport technology is first implemented in and between large cities (Haggett 1972). This implies a population base for a growing demand of transport services.

Implementing an infrastructure based on new technology means both capacity and productivity rises in the services of the physical network. Combined with a growing demand function these increases and adaptions of services run from system to system in the whole infrastructure. Intrasystem integration will work via diminished travel time per trip, will gain from other transport modes, from attraction of new travel and from increases of capacity, access and openness. Implications for railway with examples for the British Railway have been discussed by Smith (1972).

System integration work starts with speed. It aims at starting an upward circle of productivity gains and journey expansion utilising the speed of trains as the basic determinant of customer journey time. It includes margins of time delay at terminals, and margins of harmonizing meeting times with starting or the railway timetable. Vehicle speed, frequency, terminal location and accessibility, stopping patterns of trains, and punctuality are factors which constitute what is generally apprehended as journey time. Competition, mainly between air and rail transport, by means of diminished journey time leads to gains for either system but also, together with attracted new transport customers, to increased demand for end services in the integrated systems.

Names of categories in the integrated networks are suggested in Table 2. They are inspired by the terminology developed in the telecommunication discussion. This discussion also delivers the pattern of systems thinking in Table 1 and of the way integration can and is proceeding within the transport systems.

Table 2. Categories of integrated networks between transport systems and communication systems

TRANSPORT AND COMMUNICATION NETWORK CATEGORY
Passenger Transport Integrated Network (PTIN): person transport network connected by joint terminals and with time adjusted route systems.
Freight Transport Integrated Network (FTIN): goods transport network connected by joint terminals and with time adjusted route systems.
Person Communication Integrated Network (PCIN): public telenetwork with a completely integrated transmission of speech, text, picture and data.
Virtual Private Network (VPN): private network within a public tele-network.
International Virtual Private Network (IVPN): private network within a public telenetwork with a global accessibility.

Source: Forsström 1995

CTI may thus also work between the different physical and traffic networks and the communication and transport systems. Eventually, the various systems develop an integrated character, which means that time delays depending on unadjusted timetables and distance restrictions due to spatially dispersed terminals are being gradually removed. A passenger can more easily use two or three transport systems consecutively aided by computer based information systems. The distribution of information will in its turn be carried out more quickly by use of the developing various public tele- and datacommunication systems such as mobile telephones and e-mail.

CONCEPTS OF AN INFRASTRUCTURE SPATIAL THEORY

The implications of infrastructure changes on movements in the short run and land use in the long run have been noticed in this paper. Now we are dealing with substantial upgradings of the integrated infrastructure. In which way will the capacity increases and the productivity gains influence the movements, the urban form and the urban landscape as we perceive it?

In urban theory a number of region concepts have been used to delimit the sphere of urban influence. The basis have been journeys with separate destinations and the most important has been the commuting journey. A basic concept, the daily urban region is defined. It encompasses the area that is inhabited by an economically active population which for a large part commute to and work in the nodal city of this region.

However, the now developing infrastructure networks of air routes, new and up-graded railways and also motorways posess a tube quality. The passenger cannot stop the vehicle and get off on whatever place he or she wants. This banal observation has a profound implication for location.

Hägerstrand (1970) points at the differences of traveller range changes for various transport modes. The range of the car passenger is five times that of the pedestrian and his areal cover is 200 times greater. Between car drivers and air transport passengers, the range of the latter is manyfold higher but only in one dimension. The qualities of air transport have a peculiar influence on the areal range of the passenger:

"The air passenger can reach - in any case not a passenger using regular flights - not a large connected area but a group of islands, each one around every airport within reach for a back and forth flight on a single day."

"An express train, which only stops at remotely located main stations, has of course the same spatial implications." (Hägerstrand 1970, p. 4:19)

As more HST are built up, a new spatial structure of large cities is likely to emerge. The term structure is chosen instead of region with the mission to emphasize the lack of geographic contiguity between the cities.

The importance of the geographic landscape and in particular the topography is back in a new form. Before the advent of modern drained roads during the nineteenth centuary, shipping along coasts, rivers and canals, valleys with trails and a few other important trails formed a thin transport network without an areal cover. Modern road-building and the spread of railways created a dense net with an approximate areal cover. In geography the regional approach was applied to the growing urban spatial influence and the concept of the central place and region or the nodal region was defined (Christaller 1933). It became a central theoretical concept and was also much used in planning (Godlund 1954). It turned out to be a fundamental element in theory development during the positivistic era and eventually it has taken the shape of a metaphor. As such, it may work more as a steering and repeating element rather than as a renewing theoretical element.

The network term has obviously been widely adopted as a useful metaphor and approach in social sciences. It is appropriate in this context to consider its usefulness. The tube quality of the modern networks forces us to redefine the daily urban region concept. In many cases, it is no longer a region. On the regional level, it is instead a structure in space, consisting of the cities with infrastructural nodes (Forsström and Lorentzon 1991). On the local level, it includes the areal cover around these nodes constituted by a dense road and street network. By analogy, the term becomes daily urban structure. If the commuting journey is complemented by the business journey, an important new concept is identified. It is termed business urban structure.

Table 3. Redefined settlement concepts considering the tube implications of transport systems

---

INTEGRATED SETTLEMENT STRUCTURE CONCEPTS

Daily Urban Structure (DUS):

a collection of geographically dispersed cities within commuting time distance. (DUS completes the concept Daily Urban Region, DUR, which is the area extending over the commuting hinterland of an urban place.)

Business Urban Structure (BUS):

a collection of geographically dispersed metropolises within a time distance of a back and forth journey during the passing of one day. (BUS completes the concept Daily Urban Structure.)

Source: Forsström 1995

The implementation of new transport infrastructures together with CTI between transport systems leeds to a higher absolute and relative accessibility in certain places, preferably cities and metropolises. It has repercussions for the basic concept of all geography, namely space. Diminishing the over-all travel time means also diminishing the lateral distances between travel origins and destinations or a shrinking transport space.

These changes have obviously spatial nodal effects. In the short run, they tend to concentrate people and activities to specific nodes in space. As this process proceeds, the lateral distances between urban places decrease. In the long run, the formerly named "urban implosion" develops, a time threshold is passed and a new urban structure is created. It is a structure as seen from the ordinary geographic landscape point of view. The urban places of this structure can from an accessibility viewpoint be regarded as neighbouring or even contiguous and thereby make an urban space.

What about the effects of the telecommunication technology, capacity and investment development? It has been argued since long, particularly in the applied science debate, that it opens up great possibilities to reduce the need for travel (Harkness 1973, Henneman and Krzyczkowski 1974) and to decentralize human settlements and places of work (Harkness 1977). In the later scientific debate, the enthusiasm became cooler and a diversified view developed (Telekommunikasjoner 1984, IATSS 633 Project Team 1985) towards a critical view of the potential of telecommuting (Salomon 1984, 1986). Later remarks make a note of the quality of telecommunication services as being non-nodal as the telenetwork is accessible at practically every point in space (Törnqvist 1993).

It is technically true regarding mobile telephones and also practically for a coming personal communication integrated network (see Table 2, PCIN). On the other hand, the two other types of telenetworks (see Table 2, VPN and IVPN) which are created to serve business organisations are evidently their shadows. It illustrates the flexible use of teleservices as being subject to management policy.

CONCLUSIONS FROM THE SPATIAL IMPLICATIONS OF CTI, TUBE QUALITY AND TIME GEOGRAPHY

The conclusions from the mentioned arguments on transport systems integration, the tube quality of transport networks under the restrictions of time geography, can be derived for different spatial considerations. These aspects concern primarily the terminals of the network. Their function is to offer transport services. They also have a long-term local influence on location due to the areal distribution of good and bad externalities.

On the local level, the networks have generally no or little tube quality. Street and road traffic plays a dominant role. Cars are often not tied to terminals. If so, the tube effects of parking lots and of public means of transportation are to a high degree revoked by walking. Network nodes tend to coincide with the origins and destinations of transport and to make such a dense point pattern that they practically have an areal cover. The daily urban region is a justified concept on the local level. Transport network integration will strengthen its areal quality. There is a diffuse zone where the network configuration becomes less dense and gradually takes the form of a tree structure. When the web quality is replaced by a tree structure the urban region no longer has an areal shape but a linear character.

On the regional, national and international level, the transport networks have the tube quality. The integration and the technical development at this level imply considerable reductions in journey time. The transport space between the terminals of the integrated transport system is likewise diminished. Thus, the relevance of the concept time-space convergence (Dicken and Lloyd 1981) is further strengthened. It seems likely that the origin and destination nodes or the terminals will continue to cluster in a shrinking transport space.

In this development phase the importance of the tunnel-effect is crucial and the consequences for the urban structure and its image are far-reaching. The accessibility rises of the international and national transport infrastructure, which is being upgraded, enlarged and integrated, are concentrated at the terminals and the big urban nodes they serve. These rises are sometimes restricted to limited parts of the urban areas and their local transport networks. Large parts of the rest of the cities, towns and rural areas in the affected nations have absolute or relative accessibility decreases. They are not nodes in the network and the tunnel-effect makes it difficult or even prevents them, in case of day-time restricted journeys, to use the network. The cities with terminals, in this international and national transport infrastructure, are connected in a special urban structure almost without areal qualities.

The image of the time-space convergence concept has been the distorted area. The idea is that the increased node accessibility is available also in a wide zone around the node. This attempt to use distorted areas as a sign of the uneven distribution of accessibility is a mistake. The mental landscape of a migrant, which Hägerstrand mapped in 1957, is not the same as the opportunity to move within certain time limits. Accessibility is not distributed continually in space. The tunnel-effect makes the distribution discrete outside the built-up area. Day-time restrictions make it discrete from the regional to the international level. Due to the tunnel- and day-time-effects the areal conception of accessibility has to be exchanged for a tree or network image. It should also incorporate the time-distances between the urban nodes concerned. The conclusion is that an image or a map of an accessibility-based urban structure is not a common map layer ready to be superimposed onto any ordinary map.

The urban spatial influence which works via transports starts in an area, continues along lines and ends in points. By contrast, the supply of public teleservices is in principle available within areas, but at rising costs in remote parts of a region. Private activity organised networks (see Table 2, VPN and IVPN) supply influences to nodes in a dispersed or clustered point pattern. The conclusion must be that a large part of the teleservices support the transport network node localisation tendencies.

Considering the spatial aspects of the integration of communication and transport networks in an urban space, both the local and the regional conclusions have to be taken into account. The area approach of DUR (see Table 3) is obviously inadequate. An urban space of a city is a spatial subsystem consisting of a variable part of the city region in question and a number of also variable urban part areas around network terminals in superior, lateral and subordinate cities. The sample of these cities is determined by a defined time distance, which in turn depends on the purpose of the transport. In short the urban space is a structure of different city parts in transport space.

METHOD, IMAGES, DATABASES AND GIS

How to make transport space and urban space operational? The extent of transport space changes is spotted by the nodal positions. The individual nodes of transport space are conventionally measured by a number of techniques with increasing complexity, ranging from desire lines or straight distance, road distance, to time distance. The qualities of node patterns can be apprehended by topological measures based on graph theory. A careful calculation based on the fastest trip chains and including all the relevant partial trips, access and waiting times, is preferred. Time-geography puts no limits to the extension of transport space as it expresses the space between terminals.

The urban space construction may imply accessibility concepts of different kinds. The least complex concept has a simple functional relation: accessibility of a place is the sum of the quotients of the number of opportunities (activity sites) in other places and the discounted time-distances to these places. In this case the urban field is specified for a commission of some kind. The results are obtained as the summarized opportunity per unit of discounted time for all urban nodes in the integrated network.

In the urban space question, accessibility may be interpreted as a time-geographic relation. Accessibility of a place is judged by its position in a time budget in a day-time perspective. Both the commuting and the business trip can be defined in time-periods which allow time for work as well as for a business meeting. The time-geographic approach is chosen here as the primary objective to make a general accessibility framework and not a case for a specific opportunity. This definition implies that a travel time area is calculated under the restrictions of a time budget. Necessary material are air, rail and road network configurations and timetables, which are related in a database.

The travel time area is calculated for the time-distance relations of one urban place with its neighbouring places. A time-distance is accepted by the journey purpose, work or business to an average value or an interval. It is including trip time for a chain of movements starting with walking, road transport, rail or/and air travel ending by a road trip and walking. After the working or meeting period the return transport takes place. The results are in the form of end points at links in the integrated network in the origin city and in the neighbouring cities. They show the borders of areal elements of these cities, which shall make up a virtual map of urban space.

Previous attempts to show the implications are in my opinion insufficient. Mapped time-distances on ordinary maps have no visual effect beside the map itself. CAD-images in the form of topological maps are better, but can only show the time-distance impact for one place at a time. Isochrones can be used only when the tube-effect is not present. Areal distortion is represented by the transformation of an ordinary map by means of an algorithm of some kind. Examples have been based only on areal transformation. They do not take the physical properties of networks, which have been accounted for earlier, into consideration.

The work of applying these principles in a GIS is proceeding.

INFRASTRUCTURE SCHEME INVESTMENT AND DEVELOPMENT IN SWEDEN AND EUROPE

Investments in Sweden and in the European Union (EU) aim at increased transport capacity and productivity by removing barriers for trade and by implementing new technology. As Sweden and Finland have joined the EU, border-crossing will be easier. From our point of view, transaction-time between countries will decrease.

The rail and road investment programme in Sweden has expanded suddenly. Investments in railways and roads are according to the traffic policy principles of 1988 dependent of government and parliament. There is a parliament decision from 1993 of investments in railways and roads during the period 1994-2003. A sum of 98 000 million SEK has been decided upon. It includes the upgrading of the old railway network, within an area delimited by Malmö in the south, by Göteborg, Stockholm and by Sundsvall in the north, to be used by HST. The building of a road and rail bridge between Copenhagen and Malmö, at a cost of at least 17 000 million SEK, has started. A road improvement programme in the regions of Stockholm and Göteborg has started.

The telecommunication and air traffic managers and boards act on a basis of charges. Their incomes are market dependent and they make their investment decisions independently. The former Board of Swedish Telecommunication, now Telia AB, started investing in a national digital telenet in 1980. The last electro-magnetic switch was exchanged for an AXE in 1995. From 1980 onwards an airport system has been constructed in Sweden. Deregulation has not had a generally lowering effect on air traffic prices. In spite of some years of stagnating domestic air traffic, the general conclusion of Swedish public and private investment policy is that capacity and productivity will increase and that they will be accompanied by rising traffic in all sectors.

Table 4. A comparison of the heavy communication and transport investors in 1994

Telia AB	10 000 million SEK
The National Administration of Railways	6 000 million SEK
The National Administration of Roads	7 500 million SEK
The National Administration of Air Traffic	1 500 million SEK

Source: Stomnätplan 2003-2020

EU transport policy and investment support in the European infrastructure together with member nations' own efforts, including Sweden, aim at making a new and better transport and communication networks real: a European High-Speed-Train Network, a more reliable Air-Traffic-Management System and a European Motorway System. This means increased speed and thus shortened time-distances within the different systems. The policy towards common sites for airports, railway stations and bus terminals creates junctions with multiple functions. Existing or planned sites to be mentioned are Paris, Lyon, Lille, Manchester, Oslo, Stockholm and Copenhagen. The implication is coupled networks and diminished transfer time between the respective systems.

As a result of these investments and plans, there will be decreasing time-distances within and between the integrated transport systems. Increased capacity will permit even more people to travel. Travel planning is facilitated by time-table data-based information systems, which increase the access to the transport networks.

It can be concluded that the advance of global air traffic and, in USA and in EU in particular, a fractured, non-contiguous urban structure is at hand, and that some of the prerequisites for the creation of a changed air and a new rail transport space are being realized. The transport and communication space and the associated fractured urban structure of Europe will continue to develop the differentiation from the geographic landscape. A tension will appear between this fractured European urban structure and the geographic structure of the different national urban structures including business and industrial activities. These activities will adapt to new possibilities, of which accessibility opportunities are a part, and thus will gradually make new urban structures visible.

This is a small part of the efforts to work with the major geography problem of location. A contribution can be made to the sometimes very complex decision-making process of positioning an activity in space. Maps of business urban structures may shed light on the first part of this process, namely the question of localization. Maps of daily urban structures can

244 *Forsström*

contribute to solve the location question, the second part. Finally, the problem of choosing a site can benefit from this proposed geographic information system.

REFERENCES

BONNAFUS, A. (1987): 'The regional impact of the TGV', Transportation 14: 123-138.

BRAUDEL, F. (1989): Civilisationer och kapitalism 1400-1800. Gidlunds, Stockholm, 900 p.

CLAWSON, M. (1972): America´s land and its uses. John Hopkins University Press, New York.

COLLINS NEW ENGLISH DICTIONARY (1965): ed. Irvine A.H. Collins, London and Glasgow.

CHRISTALLER, W. (1933): Central Places in Southern Germany. Prentice-Hall, Englewoods Cliffs, New Jersey.

DICKEN, P. and LLOYD, P E. (1981): Modern Western Society. A Geographical Perspective on Work, Home and Well-being. Harper & Row Pub., London.

FORSSTRÖM, Å. (1995): AXE, FAX, NMT OCH TGV, in SVALLHAMMAR S. and ÅSE L-E. (eds.): Kommunikationernas Europa, SSAG,Ymer 1995, 158 p.

FORSSTRÖM, Å. and LORENTZON, S. (1991): 'Global development of communica-tion', in BRUNN, S. and LEINBACH, T. (eds.): Collapsing Space and Time: Geographic aspects of communication. Harper Collins Academic, London.

GODLUND, S. (1954): Busstrafikens framväxt och funktion i de urbana influensfälten Samhällsvetenskapliga studier No 11. Glerups , Lund.

JANELLE, D. (1969): 'Spatial reorganization: A model and a concept', Annals of the Association of American Geographers 59: 348-364.

GIANNOPOULOS, G.A. (1992): 'Innovations in urban transport and the influence on urban form. An historical review', Transport Reviews Vol. 12, No 1, 15-32.

HAGGETT, P. (1972): Geography, A Modern Synthesis. Harper, New York.

HALL, P. (1995): 'De europeiska höghastighetsbanorna och deras geografiska betydelse', in SVALLHAMMAR S. and ÅSE L-E.(ed.): Kommunikationernas Europa, SSAG,Ymer 1995, 158 p.

HARKNESS, R.C. (1974): Telecommunications substitutes for travel. U. S. Department of Commerce, Office of Telecommunications, Washington D. C.

HARKNESS, R.C. (1977): Technology Assessment of the telecommunications /transportsation interactions. Stanford Research Institute, Menlo Park, California.

HENNEMAN, S. and KRZYCZKOWSKI, R. (1974): Reducing the need for travel. Interplan Corporation, Santa Barbara, California.

HÄGERSTRAND, T. (1957): Migration and area: survey of asample of Swedish migration fields and hypothetical considerations on their genesis. Lund Studies in Geography B, Human Geography, 13, 27-158.

HÄGERSTRAND, T. (1970): 'Tidsanvändning och omgivningsstruktur', in Urbaniseringen i Sverige - en geografisk samhällsanalys. SOU 1970:14. Expertgruppen för regional samhällsanalys (ERU). Esselte, Stockholm.

IATSS 633 PROJECT RESEARCH TEAM (1985): 'Substitution/complementary relationsship between traffic and communication', International Association of Traffic and Safety Sciences, Vol. 9, 23-32.

LIVING CONDITIONS SOS (1977): Report 1, Employment and working times 1975. National Central Bureau of Statistics. Liber, Stockholm.

MATTHIESSEN, C.W. (1993): 'Scandinavian links. Changing pattern of urban growth and regional air traffic', Journal of Transport Geography, Vol. 1, No 2.

NATKIEL, R. (1970): New Society. London. From Dicken, P. and Lloyd, P.E. Modern Western Society. A Geographical Perspective on Work, Home and Well-being. Harper and row pub., London.

PUCHER, J. (1995a): 'Urban passenger transport in the United States and Europe: a comparative analysis of public policies. Part 1. Travel behaviour, urban development and utomobile use', Transport Reviews Vol. 15, No 2, 99-117.

PUCHER, J. (1995b): 'Urban passenger transport in the United States and Europe: a comparative analysis of public policies. Part 2. Public transport, overall comparasions and recommendations', Transport Reviews Vol. 15, No 3, 211-227.

SALOMON, I. (1985): 'Man and his transport behaviour. Part 1 a. Telecommuting - promises and reality', Transport Reviews, Vol. 4, No. 1, 103 -113.

SALOMON, I. (1986): 'Telecommunications and travel relationships: a review', Transport Reviews, Vol. 20A, No. 3, 223 -238.S

SAYER, S. (1992): Method in Social Science - a realist approach. Routledge, London.

SMITH, J.G. (1972):'The effects of speed improvements on the volume of traffic, as for recent years experienced by British Railways', Rail International 2:641-649.

STOMNÄTPLAN 2003-2020: Banverket 1993.

RATHERY, A. (1993): 'Traffic flow problems in Europe',Transport Reviews Vol. 13, No 1, 1-23.

TELEKOMMUNIKASJONER (1984): Nordisk komité for transportforskning, Publikasjon No 45, Oslo.

TÖRNQVIST, G. (1993): Sverige i nätverkens Europa. Liber-Hermods, Malmö.

WEGENER, M. and SPIEKERMAN, K. (1995) : Analysing the Impacts of High-Speed Train Corridors. Presentation at the Workshop 'High-Speed Train Corridors'. Jönköping International Business School, Jönköping University.

Part II

PERSPECTIVES FROM PRACTICE

14

TRANSPORT AND LOGISTIC SYSTEMS FROM VOLVO'S VIEW

Rune Svensson

ABSTRACT

Transportation and logistics have become more and more focused during the last five to ten years among all international and national industries. However the steps we have to take during the second half of the 90's to meet the logistics of the next century will be a real challenge for everybody involved.

Keywords: Volvo, common market, competitive edge

INTRODUCTION

The new demands of increasing speed to attain short delivery times with 100% reliability will change the strategies of many transportation companies. The reasons for these demands from most shippers all over the world are a high value of cargoes, the extreme customer orientation of many products, the variety of products, the increasing quality demands, more guaranteed delivery times and much stronger international competition.

For many years physical distribution costs have been the priority from shippers towards the transportation industry. This is changing and today we see more transportation systems in all modes of transportation based on a more logistical approach than ever before.

In the Volvo Group of companies we are investigating how we can meet the future demands from customers on the world market. We decided in the beginning of the 90's to reduce our lead times in all areas, design, purchasing, production and distribution by 50% in 1995.

Today we are setting new goals of another 50% to be reached hopefully in 1998-99. That means to be able to deliver a custom built car or vehicle within a fortnight to almost all places around the globe.

Two main areas must be drastically changed. We need a high degree of flexibility in our production units and on the transportation side the daily frequency to most of our international markets. This is the background to why we today are establishing more and more integrated intermodal systems based on on-line information and communication systems in order to overview all links in the transportation and production chain.

Air and shipping lines as well as rail and truck operators must be involved much more in the total concept and based on much longer contract periods than we have usually been working with in our transport relations. Ports and transfer terminals must also be efficient and faster in their turnover and probably with a higher degree of automization in loading and discharging.

We have to go back in time to find the same interest in transportation and distribution as of today. More and more CEOs and company managements are realizing what physical distribution and logistics can do to improve a company's profit, increase the share of the market, improve cash flow, open new territories and introduce new products. That is the background to why many European corporations are moving transportation and logistics from traditional traffic units on an operational or tactical level up to a more strategical level.

Trends like:

• The increasing international traffic and trade of high value cargoes;

• The need for reduction of the capital tied up in the material flow;

• The reduction of total lead times from order to final delivery;

• The higher demands of quality and reliability;

have created a whole new logistical structure for many companies and also for Volvo.

In a small country like Sweden, with an extremely high percentage of import and export, we have seen a steadily increasing number of such changes and with a very interesting result.

Since the beginning of the 80's the ten largest corporations have reduced their tied up capital by more than 60% in spite of increased production. Approximately USD 10 billion have been reduced and used for investments in new products, market activities, education and other forms to secure future development. But besides this capital rationalization, which we all thoroughly discussed during the 80's, the corporations have increased their reliability in their deliveries by approximately 70 - 75% and reduced the total lead times by 25 - 30%.

We are no doubt in the middle of an extremely interesting development, where we have found that many of the theories have become realities and the results are very interesting.

The industries of the 90's will demand and force integration between the various links in the transportation chain, and the times where shipping lines, land transport organizations and ports could work independently will be gone in the near future.

This does not mean that we all have to merge economically, but a much better cooperation and coordination in marketing, equipment, administration and planning will be a necessity. Ports and transfer terminals are vital in an integrated transportation system of today and tomorrow and must be adapted technically as well as administratively to the new world of logistics.

The economies of the industrial world are becoming more and more integrated and linked with each other. The driving forces in the world are an increase in specialization and the division of labour. At the same time competition and the fight for markets in places a long way from where the companies are situated is growing sharper. This increases the demand for speed and high quality transport.

Effective transports allow a company to compete in various levels locally, nationally and internationally. By serving large markets from few production centers transport and distribution is of strategic importance for the quality of service and deliveries.

Scandinavia has a geographical position which means that we have considerably longer distances to reach our main markets than many of our competitors from other countries. So far I feel that we have managed to overcome the handicap caused by distance by effective organization of our means of transport. However, the demands up to the next century will make us very much aware of this distance.

Consequently, to retain and develop the standard of our welfare we need to grow to a level comparable at least to that of our competitive countries. The need of increased resources, not only for our health service, care for the elderly and for education, but also in order to keep the unemployment to a minimum, and long term profitability demand that we create conditions for such growth. One of the conditions is a well designed and functioning infrastructure.

DEVELOPMENT OF THE INFRASTRUCTURE

Europe's, United States' and Japan's position as highly industrial areas and their high standard of living have been closely connected with the highly developed infrastructure. In the old days the trade and production were concentrated around ports and waterways, but the big step in the industrialization, especially in Europe and United States, took place when the railway network was built during the end of the 19th century and the beginning of the 20th century.

Fifty years ago roads and truck transportation made it possible to develop the next step in the present industrial structure. During the 50's and the 60's production was focused and transportation was handled in a traditional way in spite of that unitization mainly by the use of containers and trailers started its revolution.

Since the second half of the 70's and through a stronger competition from our colleagues in Japan and other areas in the Far East more and more of the industrial world has realized what transportation, or in a wider perspective logistics, offers to a modern organization.

LOGISTICS AND CHANGES IN CARGO VALUES

The change from low to high value products, the need to increase the turnover of capital and the new customer oriented marketing made a drastic change in valuating modern transportation.

Production logistics based on the Just-in-time philosophy or similar formulas are changing the material flow to highly frequent and punctual deliveries and started mainly from supplier to the assembly or the production line. Robotization, and what we today call flexible manufacturing system (FMS), increase the production flexibility and make it possible to develop much greater product variance.

Marketing logistics offer high quality deliveries and increase the competitiveness. This starts to play an interesting role for the rest of the 90's.

Communication logistics integrate manufacturers and suppliers, but also all other links in the total transportation chain by transaction processing systems mainly through the use of electronic data in the chain linkage.

We are in other words facing a new industrial step where transportation and logistics will play a vital role for successful development of the companies, but also for the country in total. However, all this is depending on an upgrading of the infrastructure network, roads, railroads, ports, airport terminals and of course the communication and information systems.

We have also during the latest 15 to 20 years seen a steady increasing value per ton cargo, which in reality means a change in the transport pattern. With the number of variants in our production, the high value for inventory and the necessary increase of the turnover of tied up capital will demand a much better transport efficiency than we have today.

In our history we have many examples that the most refined and high valued cargoes in every epoch first move by the most sophisticated mode of transport at that time. The further development of the infrastructure helped more and more of the low valued cargoes to have access to these systems.

One of the examples is during the 19th century the optimum technique of the sailing ships - clipper ships - the most beautiful ones ever seen on the seven seas. They were probably the most complete sailing machines and very fast and by then able to attract the high value cargoes of that time on the transocean trades. However, in spite of the optimum of sailing technique they were sentenced to death, as the new technique - steam engines - took over very fast.

A comprehensive study based on material from US, Scandinavia and other countries shows that choosing the mode of transportation is primarily based on the cargo value. 10% higher cargo value in United States leads to 2% increase of the trucking business and 1% reduction of railway transportation.

VOLVO'S MATERIAL FLOW

Volvo has a specific need to further increase the frequencies and speed to have our material from supplies to production as well as our products to our customers all over the world.

In comparison with many of our colleagues in the automotive industry our suppliers are mainly based outside Sweden and 90% of our customers are also outside our home market. Our geographical location, long distances within Sweden and above mentioned factors give us a difficult distance handicap.

We have during the years tried to compensate this by developing efficient transportation systems. This is basically the background to why transportation and logistics play a more and more important role in our overall strategy at Volvo. At the same time more and more of our production is, as mentioned before, built towards a customer order situation. All units, cars, trucks, buses etc. are more or less tailor made for a specific customer. When a customer signed an order we had during the 80's a delivery time of three to four months. We have intensively worked to reduce this and are today able to deliver within a month for most of our normal products. The goal is within a couple of years to be able to reach a time from order to delivery of approximately two weeks to our main markets in Europe and North America. To other transocean markets we probably have to add another week. The reliability in our future transportation and logistic systems is therefore extremely important.

During the 80's and the first half of the 90's we have reached a high logistical level, based on present modes of transport, existing operational techniques and present infrastructure. However, with the technique of today we still have good possibilities to develop and further sophisticate our logistical systems, based on the new IT-technology, which gives a much better communication and information system on-line, both for the inbound material flow and the export of our products.

INCREASE OF TRAFFIC

Since the end of the second world war we have seen a steady increase of the traffic volumes both at the national and the international level. From the mid 70's, the traffic volumes in most countries have grown at the rate of two times the GNP growth.

We have all reasons to plan for the same increase up to year 2000, which in many countries means almost twice as much traffic to handle as we have today. A number of questions arise.

- Do we have plans how to handle this?

- What mode of transport will handle most of these increases?

- Will the European infrastructure cope with this development?

Unfortunately most of the European countries have not been able to build up the infrastructure resources to meet this increase and the risk is that we will face a number of problems, basically in the big cities.

For the industrial development, especially based on the logistical systems, this is an unfortunate development.

Somebody has mentioned that the speed in central London is today exactly the same as we had in the beginning of the 19th century, when horses and coaches were the modes of transportation.

Most people realize today that we have to spend much effort and activity in order to find new ways of handling mainly the passenger traffic in the large urbanized areas. Unfortunately we have not up until now found the key to the new method of dealing with this matter, though hopefully we will see some innovation during the rest of the 90's.

TRANSPORT SYSTEMS OF THE 90'S

In the industrial development in the future very much will be depending on the efficiency of the transportation and the logistic activities. Especially in the new Europe of 1992 with a much bigger market for many companies there will be a greater pressure on the transportation companies to offer complete systems door-to-door or floor-to-floor connected with on-line communication and information systems. High frequencies, punctuality and reliable handling will be the key words for a successful operation.

To build up sophisticated total systems the transportation companies must be of a rather good size and we are seeing all over the world a number of mergers, joint ventures and closer cooperations between the companies in transportation.

Earlier we had shipping lines merging, arlines merging but more and more we find mergers or cooperations between different modes of transportation into what we may call "transport warehouses". With the present trends and activities going on we will have a rather fast development of this new form of organization and probably all over the world.

Many companies, mainly small and medium sized units, see a risk in these big transport organizations, but these organizations will probably offer a much more system oriented operation, which no doubt will help such small and medium sized units to find a better way to reach their markets or to buy components from far distant suppliers.

In the relations between shippers and transportation companies there will be a much better integration, specially within the communication systems.

The first steps have already been taken with the bigger transport organizations, where the documentation and communication including the payment procedures are all electronically handled. That means of course also a further drastic reduction in the still existing documentation which is still used in transportation, and of course the great possibilities of further rationalization.

MODES OF TRANSPORTATION

The logistical concept has increased the road transportation much more than what has happened in shipping and railway. 1993 we had more than 160 million vehicles in Europe. The amount for the whole world was approx. 510 million vehicles, of which approx. 25% consisted of trucks and buses. In spite of the heavy subsidies to the railways the trend towards more road transportation is steadily and clearly upward.

The general increase among the top-twenty industrial countries in the world is more than 30% for the last ten years. For western Europe it has been slightly less, approx. 25%. The strong economical increase in southern Europe has showed an increase of the road transportation by more than 50% and compared with a country like Sweden we have during the same period stayed at 15%.

Even though the cancellation of borders and the custom controls within Europe has become a fact, the increase of volumes has filled up the extra capacity for road traffic and no doubt we will be facing certain congestions around in a number of areas, mainly in the northern part of Europe.

Regarding the rail transportation the general increase of volumes among the twenty industrial countries in Europe has stayed around 7%. In more specific countries it has been higher, but in certain cases like Japan there has been a drastic reduction. The United States have been able to keep their development with approx. 7% increase of rail road traffic. That means that the rail road market share has since the mid 1960's been reduced from approx. 33% to 20% of today.

DISTANCE TRANSPORTATION WITHIN THE COMMON MARKET

Table 1. Market shares in percent

Mode of transportation	1965	1970	1975	1980	1985	1990	1995
Road transports	48	52	59	64	67	69	71
Railway and inland water transports	52	48	41	36	33	31	29

These figures show that we are moving away from big volumes on slow going modes of transportation to highly frequent, flexible and small transport units. Logistics is therefore one of the main reasons why road transportation has had such an enormous growth until now all over Europe.

In the future we can see the same trend coming into air freight, but presently the airlines have big problems to handle their traffic in the air and also of course a lack of space in the air terminals.

However, for the long and regular big movements of bulk, mineral and heavy loads, the railroads will have a potential and an important role to play during the rest of the 90's. Another good potential for the railroads will be the combi traffic and of course block trains between the main terminal points. New systems of a better combination between railroad and trucking companies must be achieved in order to offer the industries a total door-to-door concept.

The magnetic level rail concept as in the Maglev projects in West Germany and Japan is very interesting, but will not within the next ten to fifteen years have any effect on the cargo transportation, unfortunately.

THE NEW INFORMATION AND COMMUNICATION SYSTEMS

In the last few years we have become increasingly conscious of the enormous existing potential in this field. More information and communication systems will be the next milestone in the development of the shipping and transportation business. Already in 1986 we took the first step towards the on-line systems, where the different links in the transportation chain are integrated and where the complete control of the material flow can be carried out from the supplier to the final user.

Corresponding systems have also been built for our outgoing traffic, or finished products to the final consignee. This naturally provides extremely interesting perspectives and possibilities both to influence even further capital investment and increase the degree of service within marketing.

The formation of communication systems like this will influence a large proportion of our development during the 90's and primarily give us the early warning systems we have been waiting for so long. The subsequent development phase naturally includes rationalization of the multitude of documents and I think that we already during the years to come will be getting more work done with the aim towards document-free transport.

Since the technique for information transfer from mobile units exists, there is no problem in integrating railway and road vehicles as well as ships in the transportation chain in ways which differ greatly from those of today. This trend naturally makes new demands on the organization form within the transportation companies where typing out and handling of documents still make up a large proportion of dispatch work. In any country, wherever you are, wherever you live, it is impossible to stay out of information in the future. You can always run from information, but never hide.

THE TRENDS

Based on the present situation we have a fairly reliable scenario for the rest of the 90's. There are no foreseen stumbling blocks, which can change this in a drastic way, which means that we also have a good plan for our investments in logistics and transportation.

There are three overall basic trends:

- The need for expansion and the high grade of the network of motor ways to meet the demands of the increasing number of vehicles;
- The adoption and building up of an on-line computer communication network both for cargo movements and passenger traffic;
- A further development of a frequent network of air connections.

Beside these three and as mentioned before, we will no doubt see good development of the high speed trains between the main cities in order to handle the heavy passenger flow between urbanized cities.

In the beginning of this century the average person in Europe travelled between 500 - 700 metres per day. Today the average travelling distance is 75,000 metres per day and person. 75% travels by car.

Many people may have a certain reaction towards trucks and cars, especially as they have a certain influence on the environment. There is however a strong belief and a tremendous activity to reduce the risks and we believe strongly that we at the end of this century will have no big problems with the traffic from this point of view, with the exception of some very specific populated places, mainly the bigger cities.

The car is still the cheapest product per kilo in the world, in comparison with the high technical equipment in the product, and it is still the best tool for human beings. Big changes have taken place in the automotive industry from the 70's and the 80's and 23 automobile companies do not exist any longer, and more mergers and cooperations can be expected.

No doubt working with logistics the coming years will not only be an exciting time of changes, but also for most companies a complicated situation. In Europe as an example a new growing EC will harmonize their rules and regulations, which will make travelling and trade simple within the countries, but puts a rather strong pressure on the transportation industry to meet the new demands of logistics. These changes will have a tremendous impact on the inter-European traffic and have a drastic effect on the building of the infrastructure in Europe towards the next century.

In our strategy for logistics we have also a number of trends, which we have to adopt in our new transportation strategy:

− Changes in the production technology;

− New information and communication systems;

− Increased degree of service;

− Increased production and distribution quality;

− Necessary capital rationalization;

− New legislation, rules and regulations;

− New organisational patterns.

There are still many people in logistics who are focusing on the physical distribution cost and specifically the freight rate. More and more however consider the overall view with inventory, buffer stocks, reliability and punctuality, which together will be instruments both for rationalization and marketing.

Against this background it is essential to coordinate the policies adopted for business and traffic within the framework of a country's business policy. Efforts to strengthen the industry could be in vain if its requirement for effective transport is not met.

For example Sweden's industry is particularly vulnerable vis-à-vis its international competitors:

- Its peripheral situation with regard to many of the future markets of Swedish products;
- Its own small transport market, which undermines the basis for a high frequent direct transocean traffic;
- Imbalances in the domestic and internal flow of goods with capacity poorly utilized as a result;
- Uncertainty as regards future traffic policy decisions in a number of respects for example investments in infrastructure, duties and taxes, etc.

Some of the above mentioned circumstances create relatively high transportation costs for the Swedish industry and business. It is therefore important that national and local policies contain formal and more long term measures in respect of industry and traffic.

CHANGES IN THE TRANSPORT BUSINESS

Increased demands for quality in production bring increased demands for quality of distribution in their wake. Transport systems offering high frequency deliveries, better precision and shorter lead times will be necessary, partly for the work of rationalization but also as an important marketing aid.

This sort of development has radically altered traditional transport methods. As an example we have been able to compete against considerably larger producers thanks to increased automation and the use of robots. We have been forced to become more customer oriented and to refine our products.

As is well known we have high technology products. This in turn means an increasing value, which alters the pattern of transport, stock keeping and of course the location of warehouses for goods for which there must be a buffer stock.

These new demands on the pattern of transport have not attracted sufficient attention, with the result that when we were building up our infrastructure in the 80's, our investments went in the wrong direction. Unfortunately we try to keep the traditional methods of transport alive in many areas of Sweden and investments are still being made in parts of the infrastructure, which should perhaps be reduced or axed.

A typical example is that for many years we built roads, not in order to cope with increasing traffic, but to provide a means of employment.

Quantities diminish for every delivery, which is why we are moving over from large scale to small scale transport solutions. This is also why air and road transports are increasing rapidly at the expense of shipping and railways.

LOGISTICS - THE COMPETITIVE EDGE

During the 80's we have in most corporations worked with logistics as a tool for capital rationalization. The change from low to high value products in most industrial countries, the high interests and increasing inflation stress the need to increase the turnover of capital within the industry.

For the transportation industry the first effect of logistics was a rather high pressure on the physical distribution costs, mainly the freight rates. That is why probably many of the shipping and trucking lines, railway companies and airlines did not find logistics the most interesting way of developing transportation systems. However, all over the world we see the change within many companies where the traditional shipping and traffic functions are organized into a logistical department or division.

At the same time we realize that during the rest of the 90's logistics will more and more be considered as an important marketing tool. Many industries - and you will find it very common in the automotive industry - are moving towards an extreme customer oriented production. Logistics will no doubt be one of the tools to obtain competitive advantages.

There are many reasons, arguments and ideas behind this focusing on logistics:

- Transport costs have increased more than most other factors affecting the final price of the commodities. However, taking into consideration the present situation among shipping lines and other transport companies we will probably have to face further increases in the physical distribution costs during the rest of the 90's.

- Transport efficiency influences to a great extent the inventory of buffer stocks at different points of the trajectory of a commodity from raw material to the consumer. Inventory of various stocks does not improve commodity quality and also, as mentioned before, the amount tied up in inventory and distribution usually makes up more than 50% of the tied up capital. We will still find a good rationalization potential in this area up to year 2000.

- In the information society the distribution system will become an increasingly important competitive factor.

- High quality distribution, including delivery, safety, frequency and speed, will create possibilities for increased customer adaptation and the high degree of service will subsequently increase turnover possibilities.

- The increasing variation in the assortment of products results in the impossibility of keeping a high level of stock at the retail level. In order to retain and maybe even promote a high degree of service, it is necessary to build a new expanded communication network with improved wholesalers and suppliers contact preferably through a line system.

In our strategies we have to take into consideration certain basic conditions in our environment concerning our planning of the distribution systems of tomorrow. We must take into consideration:

- Continued mixed economy with relatively high taxation of vehicles, fuel and operation costs;

- Continued bureaucracy concerning permits, fees, rules and regulations etc.;

- More strict control that rules and regulations are observed. This also includes considerable penalty for violations;

- Continued contribution systems within certain branches of transport and also for certain geographical areas;

– An increased quality concept including premiums within the transport business, permitting profitability for transport companies investing in systems development.

This is the background we are working with at Volvo Transport Corporation as a base for the transport and logistics strategy and I do not think this deviates from the demands, wishes and requirements within other branches and other companies.

THE CHANGE IN THE ORGANIZATIONAL PATTERN WITHIN TRANSPORTATION

The different alternatives concerning the same marketing area from the 60's and the 70's have changed into firm coordination and a limited number of transport businesses.

Within the shipping- and haulage branches we are approaching a model, which in part has existed within the air lines and railway business, that is to say fewer dominating organizations and in many countries between one and two units.

Many people consider this to be a great risk, others see it as a necessity in order to cope with the international competition. In the world of transport in which we lived, this development was not particularly desirable. The primary objective was of course the price and service competition through the number of different alternatives.

Today the systems developed demand large transport organizations with the power, volume and financial background to manifest themselves internationally. This leads to an increasing number of coordinations for better or for worse.

In the nordic area we have seen the trend very clearly. For example in the North Sea traffic, where we only six years ago had seven to eight shipping lines, today we have one dominating one and one other alternative.

In the transocean shipping we are moving more and more towards fewer shipping lines, and maybe within a couple of years we will see, at least in Scandinavia, one or maybe two dominating transocean liner companies; some larger numbers to and from the continent but the number of alternatives heavily reduced.

The same trend exists also in forwarding, where today there is a rather considerable coordination and we will probably end up with a limited number of organizations within this field.

The times imply even further coordination, not only within the individual transport branches, but also between them. One future pattern will consist of a limited number of large companies offering total systems.

Besides this we will naturally see a number of specialists offering their services to the market or the transportation contractors. They will undoubtedly be successful and service minded in different branches of the transport chain, but their growth will be limited.

LOGISTIC SYSTEMS OF THE 90'S

As mentioned the big evolution will no doubt be on the communication and information side of the logistic system. The first step, which we are in the middle of, is to integrate the different links in the total logistic chain, like terminals, ports, custom authorities, and that will be a perfect tool in the industries' planning, marketing and financial planning.

The next step, which we will see within the years to come, will finally be that all documents used in transportation will be reduced and replaced by an on-line electronic communication system.

This trend will also make it necessary for governments and authorities to plan for a good infrastructure. As logistics will increase the use of small scale distribution systems, we will probably see more traffic moving over from the traditional big scale systems like rail, barges, river traffic and therefore the need for a good road network is essential.

In spite of all the problems in Europe the number of investments in new tunnels, roads and also some railway links for high speed trains is enormous. With regard to the environment problems we are working hard, especially in the automotive industry, to avoid these problems which partly are generated from traffic. There are great hopes that we at the end of the 90's have been able to develop engines and different types of catalytic converters, which will make the traffic a fairly "clean" operation.

With the transport volume increase in the industrial countries of two to three times the GNP growth per year, logistics will be of great importance to all corporations in the 90's. The stronger international competition will make it necessary to take full advantage of logistics as a powerful competitive weapon.

By organizing, focusing and pushing logistics and transportation towards excellence, companies can gain competitive advantage. That is why transportation and shipping divisions, based on a logistical strategy, will play a vital role for the development of any company.

The need to participate in the strategical planning is essential to reach success and contribute to a good profit and a fine development for the rest of the 90's.

15

SKF'S EUROPEAN TRANSPORT NETWORK: A CASE STUDY

Henrik Lindeberg

ABSTRACT

During the eighties the SKF distribution structure in Europe consisted of seven international warehouses and a local warehouse for each domestic market. The international warehouses were situated in Gothenburg, Katrineholm, Luton, Schweinfurt, Airasca, St Cyr and Spain. The decision was to implement a totally new distribution structure called NEDS, New European Distribution Structure. The NEDS project is leading to a restructuring of both the warehouses in Europe and the transport system. When a system like this is operated, there is a big need for information about many conditions and states. One of the big parts in NEDS was to develop a new system platform to manage the new information flow.

Keywords: SKF, distribution structure, Scansped

BACKGROUND

During the eighties the SKF distribution structure in Europe consisted of seven international warehouses and a local warehouse for each domestic market. The international warehouses were situated in Gothenburg, Katrineholm, Luton, Schweinfurt, Airasca, St Cyr and Spain.

This structure for distribution of SKF products had been developed during many years. The way the structure looked and worked was a natural development of the manufacturing strategy. All the international warehouses were placed close to a production unit.

Figure 1. Previous structure: international warehouses

In total this meant that there were 25 warehouses all over Europe providing service to SKF customers. The service to the local market in each country was provided from the domestic warehouse and the assortment in this unit should cover a large part of the total SKF product range. This was considered a good distribution system, where the local warehouse personnel had good knowledge of the customers and were able to provide a direct service of products to the customer. When the domestic warehouse needed new products or products not available from their stock, an order was placed in the international warehouse that should store the product. The system was rather slow, and when a new product was ordered, it could take quite a long time before it arrived.

All these different storage points in Europe meant that the total value of the inventory stored was very high. This was necessary to be able to provide a good service to the customer. To reach the customer there were at least three storage points involved. First the factory stock, then the international stock and the domestic stock. The distribution structure with the seven international warehouses and eighteen domestic warehouses is described below.

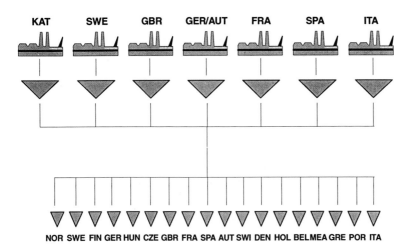

Figure 2. Previous distribution structure

Even though the domestic stock covered a large range of the total assortment there were always situations when the product wasn't in store or available in the domestic warehouse where it was requested. This situation meant that SKF was not able to satisfy the increasing demands from the customers.

NEW EUROPEAN DISTRIBUTION STRUCTURE - NEDS

In the beginning of the nineties a totally new distribution structure was discussed. Due to increasing demands on speed and frequency from the customers something had to be done. At the same time the board and management of SKF started to address the problem of a too high inventory level. With the old distribution structure the only way to provide better service was to increase the stock or start to do more frequent transports of goods. To increase the stock would not solve the basic problem and to increase the frequency within the old system would not be cost efficient. The question to answer now was how to provide a better and more frequent service and at the same time reduce the stock significantly.

The decision was to implement a totally new distribution structure called NEDS, New European Distribution Structure. The NEDS project is leading to a restructuring of both the warehouses in Europe and the transport system.

The seven international warehouses should be reduced to four. A new European Distribution Centre, EDC should be built to provide service to the aftermarket and small customers. This new warehouse was to be situated in Tongeren, Belgium.

Figure 3. New European distribution structure, NEDS, international warehouses

The new approach was that the European Distribution Centre should provide a daily service to the aftermarket and small customers all over Europe.

With this system the products could reach the customer in basically three ways:

1. Direct Customer Deliveries (DCD) from the factory or the international warehouse;

2. Deliveries to large customers from the international warehouse;

3. Deliveries to the aftermarket and small customers from the European Distribution Centre (EDC).

Figure 4 describes the new distribution structure with four international warehouses and a European Distribution Centre. It also shows how the products reach the customer through different distribution channels.

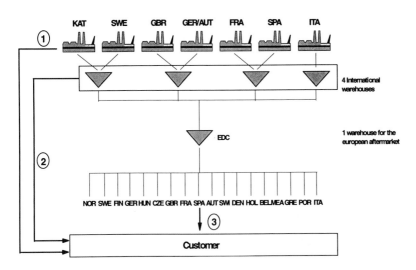

Figure 4. New European distribution structure

The new distribution stategy has a number advantages compared to the old structure. The reduction of number of storage points gives:

- reduced total safety stock;
- improved total service;
- reduced total operational costs;
- improved assortment control;
- quicker response time between market and production.

THE OLD TRANSPORT NETWORK

In the old distribution structure the transports were based on deliveries with full trucks. The trucks were driven between all storage and production sites in Europe. This meant that there were a large number of relations. In order to get this kind of transport system economically favourable, the trucks were sent away when the volume reached a full truck load. This system was really cost efficient and for some of the relations in this network there was a full truck every day, but for others there could be just one truck per week or worse. This meant that the service provided by this system couldn't satisfy the new requirements from the customers. A picture of the old transport system structure can be seen below.

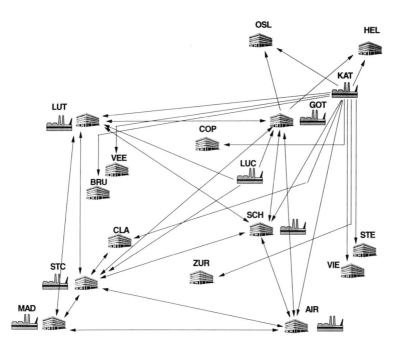

Figure 6. Old transport structure, factories and warehouses

The total transport time in this system is also long mainly due to the low frequency of transport to some markets. This caused some problems when the customer needed the product quickly. This situation happened quite often in this old system and the cost for these transports were quite a large part of the total cost.

DAILY TRANSPORT SYSTEM - DTS

In order to create a new transport system that should fulfil the requirement of the customers a totally new transport structure had to be developed. The new transport system DTS is based on daily transports in Europe. To make this cost efficient the old way of transports between all production and storage sites had to be changed.

If the old system had been used with daily transports, this would have meant that many trucks only would have transported a small amount of goods. The new system looks and works very similar to the systems used by the big express delivery companies.

To get an acceptable loading degree in the system, there had to be groupage of goods designated for different places on one truck. Then the truck had to go to a hub somewhere in Europe where the goods could be reloaded to a new truck heading in the direction of the final destination.

Figure 7. New structure: warehouses and international hubs

In the new transport system there are four international hubs in Europe. They are situated in Gothenburg, Schweinfurt, Tours and Ede, see Figure 7.

The service provided from the European Distribution Centre to the aftermarket could be described as follows:

- Central Europe (Germany, Belgium, Netherlands, North of France and parts of Switzerland) is covered next day;

- The rest of Europe is covered in two to four days depending on transport distance and frequency.

To make this possible and reliable a schedule with fixed times for departure and arrival to different hubs and destinations had to be developed.

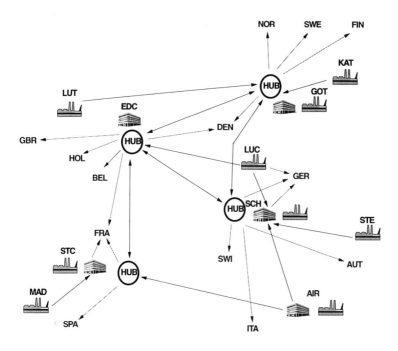

Figure 8. Daily transport system, HUBS, factories and warehouses

The goods are sent out from the shipping unit on a truck heading for a hub. When the goods arrive to the hub at a specified time the goods are unloaded. The transport packages are scanned and checked to make sure the truck was complete. Then the goods are sorted according to destination. The truck for the other destinations will leave after a fixed time schedule. The transport packages will then be scanned again and loaded on the new truck heading for the next destination. All this is done to make sure that all the received goods at the hub will be put on the correct truck and that the shipment is complete. This working procedure is necessary to keep track of the goods and make sure that everything is done correctly. When the goods have arrived to the final destination the domestic hub or receiving part can check the content of the truck by reading the packaging specification and compare it with what they were expecting.

Figure 9. Example of transportation via international hubs

IS-SYSTEM DESCRIPTION

When a system like this is operated, there is a big need for information about many conditions and states. One of the big parts in NEDS was to develop a new system platform to manage the new information flow. The Daily Transport System is operated by Scansped with no interference from SKF. SKF only specifies the requirements and Scansped tries to fulfil them in the best possible way. The requirement and specifications were worked out in cooperation between SKF and Scansped.

SKF provides a file with information needed by Scansped to operate the transport system. The file is called a pack specification file and is sent from the ICSS or SCSS system at the warehouse. ICSS and SCSS are order handling systems used in the warehouses in Europe. The international warehouses use the ICSS system and EDC uses the SCSS system. These two systems are containing the information about the orders that are going to be shipped to the customers. Figure 10 shows how the different systems are connected to each other, to customer factories and suppliers.

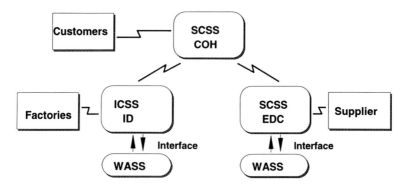

Figure 10. SKF ICSS and SCSS connections

The pack specification file is sent by MEST or MEMO to Scansped and contains the following information:

1. Supplier identification;

2. Warehouse Code where the goods are sent from;

3. Load Set Number;

4. Customer Number;

5. Invoice Number;

6. Transport Mode Code; what kind of transport is to be used;

7. Vessel Name. Trip Number;

8. Delivery Adress;

9. Transport Packages Numbers;

10. Gross weight of the Transport Packages.

Figure 11 gives an extract from the packaging specification file. It is quite a small file containing the information needed by Scansped.

File transmission from SKF
Extract from Packspec. File

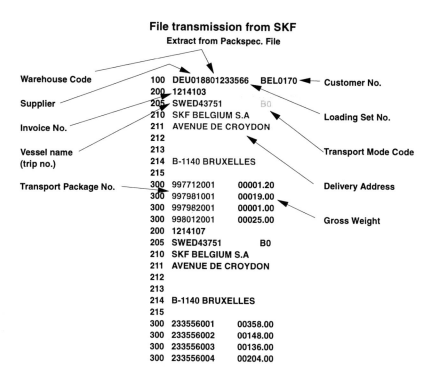

Figure 11. Extract from the SKF Packspecification file

This information is sent to Scansped from the SKF shipping units via a mainframe computer at SKF in Gothenburg. From this point Scansped takes care of the data and uses it without any interference from SKF.

Scansped sends the information to their different transloading hubs. By doing this the hubs will know what goods they can expect to arrive. When the truck arrives to the hub the transport packages are scanned. If there are any deviations in the goods received compared to the information provided by the system an error list is presented. This information is also used to show when all the goods for other destinations have arrived. When the new truck is loaded before it is heading for the new destination all transport packages are scanned to make sure the truck leaving is complete with all the goods.

To keep track of the goods and to be able to provide statistical figures, Scansped manages a tracking and tracing system called CIEL. This system has an on-line connection with some users within SKF and can be used by them to see where the goods are for the moment.

By using the CIEL system Scansped is able to gather historical data about all their transports for a specific time period. This is used in follow-ups like making sure that the trucks arrive as it was specified in the time table.

As mentioned above the pack specification file also contains some basic information about the transport packages and their weight. This is used to specify what the different customers should pay. The cost for an SKF company is based on tonkm transported to make the payment of the bill more manageable. Otherwise there should have been a lot of quite complicated calculations to see how much each SKF company should pay. The payments are done in the TMS system which is the Treasury System in SKF. This system is connected to the central computer of Scansped. Scansped sends a specification of the costs, weight transported, number of trucks, etc. each week to SKF to make it possible to do a follow-up.

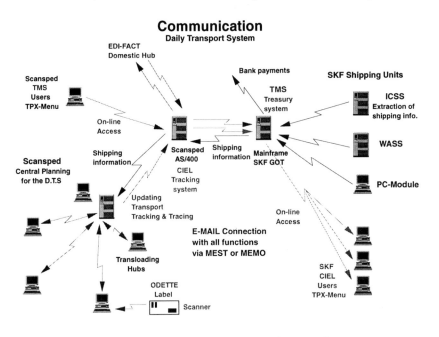

Figure 12. The communication network between SKF and Scansped

SUMMARY

The new European Distribution Structure, NEDS is an answer to the increasing demands from the customers. The concept is a way to provide a better and faster service to the customer of SKF at a reasonable price level. It has also made it possible for SKF to decrease the total inventory level. The NEDS project is now almost finished and the benefits of the new system can clearly be seen. The total stock level has decreased from 30% to 25% of net sales. The total transport cost with the new system with much more frequent transports is unchanged compared to the old system. One big reason for this is that the special express shipments were very common in the old system. The two main parts in the project, the warehouse restructuring and the implementation of a new transport system are depending on each other. If the old transport system should still have been in use, the new structure of the warehouses would not work.

The new distribution structure is depending on an IS-platform that suits the needs of the customer, SKF and Scansped. This system for distribution of SKF products works really well today and we think that it will provide sufficient service for a couple of more years. How big the changes are that have to be done then is depending on the market. We have to listen to the customers and try to provide the service and products satisfying their needs.

16

ADVANCED LOGISTICS IN EUROPE: THE CASE OF BTL GROUP

Bo Ireståhl
Thomas Kanflo

ABSTRACT

This paper describes the strategies and change process of the European part of the BTL Group in response to the rapidly changing market situation in Europe. The focus is on the development of the physical transport system, the development of information systems and the integration of these systems into a functioning and efficient logistics platform that will strengthen the market position of the Scansped Group.

Firstly a short presentation is given of the BTL Group, secondly we discuss the changes in the European market situation, thirdly we present a theoretical framework that tries to explain our development processes and fourthly we present the future strategy of the Scansped Group, the roles of the FUTURA transport system and the CIEL information system.

Keywords: BTL, Scansped, market changes

THE BTL GROUP

The BTL Group is a European based transport and forwarding company. We are operating at 500 different locations in 31 countries, employ about 11000 vehicle units transporting some

18 million tons of freight each year. The number of employees is 10,000. In additon to these figures another 5,000 persons are working with the collaborative haulage contractors and 5,000 at associated companies. The net turnover in 1995 was SEK 16 billions ($2.2 billions). The Group is divided into four divisions, working with different geographical areas and/or different products.

- Division Scansped Group works with land transports in Europe;
- Division Bilspedition Sweden works with domestic land transports in Sweden;
- Division Wilson Group works with air and sea transports on the global market;
- Division Specialist Companies is a group of small transport companies, working in certain niches, e.g. transport of hanging garments, porcelain, furniture etc.

THE SCANSPED GROUP

The European logistics activities of the BTL Group are gathered in the Scansped Group which consists of 27 sudsidiary and associated companies with 270 offices in 17 European countries. The number of employees are 4,400 at subsidiaries and 3,600 at associated companies. The net invoicing was in 1995 SEK 6.3 billions.

The Scansped Group employ around 7,000 load carriers to move freight on more than 2,000 scheduled routes.

The business concept of the Scansped Group is:

"Scansped will develop and produce transport, logistics and information services which meet the market's demand for quality, efficiency, simplicity and environmental responsibility".

The Scansped Group is currently in the middle of a very large change process, aimed at increasing the performance on the changing European market. The change is intended to result in a more efficient European transport network taking full advantage of an integrated information system with complete EDI support.

The ultimate goal of the change is of course to meet the demands of the changing market and thereby increase our competitiveness and strengthen our market position.

CHANGES IN THE EUROPEAN MARKET

The European transport and logistics market is undergoing substantial changes. The most significant change is of course the creation of the SEM, the Single European Market, within the Europen Union (EU). Since January 1st 1995 Sweden is a member of the EU. The changes in the European market mean that we have had to get used to many new aspects of business. First of all we think that the changes will mean that the need for transport and logistics services will continue to increase as deregulation in trading increases the goods flows between different European regions. Further, the market is gradually being deregulated in some areas, e.g. less formalities with customs and border crossing activities.

In some respects however the market is being more regulated, e.g. regarding the size and length of vehicles and trailers. It is also obvious that the competition will increase, especially when we reach the point when domestic transport in a third country is allowed.

In addition to the creation of the SEM the opening of the previously closed countries in the eastern part of Europe means a completely new market for logistics providers, as well as the occurrence of new competitors who are today producing transport services at low cost but are perhaps less logistically advanced. The BTL Group is certainly aware of these new markets and has already approached many of them, starting new transport routes to most of them, and setting up subsidiaries or partner relations in many countries.

Another important change, that we think would have happened even without the SEM, is however the changes that can be noticed in customer demands in the existing markets. These changes have been discussed by researchers and business representatives for several years but it is not until recent years that they have actually begun to appear in the relations between client and forwarder.

Some of the most important customer wishes are today:

- Increase in frequency with reduced quantities in each shipment;

- More advanced logistics services;

- Monitoring of goods;

- Increased transport quality and reliability;

- Fewer suppliers, also regarding forwarders and transporters;

- Long-term relationships.

The above presentation of changes is of course very brief and not intended to cover everything that is happening. In this paper we however feel that the subjects discussed are sufficient in order to make the reader understand the strategies discussed in the next sections of the paper.

A THEORETICAL FRAMEWORK FOR THE CHANGE PROCESS

From a theoretical point of view the change process that we are in can be described using the model of Figure 1.

The most basic goal in our business is of course to make our clients, by which we mean both the shipper and the consignee, satisfied. If we cannot give the clients what they want, there is no point for us in being in the business at all. A client's demands are really not very complicated; basically he wants his goods picked up and delivered with the level of transport quality agreed by him and us.

Apart from this goal the purpose of being in the business is of course to make a profit. The profit is required by the owners, who want return on their invested capital, and they normally also want to secure the continuous well-being of the company in order to show future profits. A certain level of profit is also wanted by several other groups, e.g. the employees who want to keep their jobs and increase their salary, the clients that need our transport network and the government who realises the importance of a well functioning transport system.

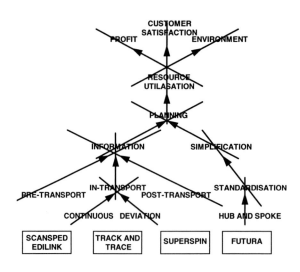

Figure 1. Model for development of Scansped Group, with examples of concepts and IT applications

In addition to this we have at the BTL Group a desire to be as environmentally aware as we possibly can. As we think it is an inevitable fact that transport companies must exist both today and in the future, at least as long as the structure of the society looks even the slightest like it does today, the thing for us to do is to fulfill the needs of the society using the least amount of resources possible. As we will show below this is very much in line with what we try to do also in reaching the two previously discussed goals, customer satisfaction and profit.

With the goals stated above, the most important focus has to be resource utilisation. By this we mean of course the utilisation of trucks and trailers, but also infrastructure such as terminals, roads and rail, computer systems and not least personnel. We find that a high utilisation of resources is the factor that without any competition has the largest positive impact on both profit and the environment. The explanation of the importance of resource utilisation is really simple: Using our resources optimally reduces the need for resources, and fewer resources need less capital, both tied in the resources and in the operation of the resources, and very important, fewer resources means less abuse of the environment, both while creating the resources and while using them.

This reasoning leads us to the fact that resource utilisation is our key operational variable, and that our strategy has to aim at optimising this variable at all times. The secret of resource utilisation is planning. The better we can plan the usage of resources, the better the degree of utilisation will be. We see two main ways to enhance the planning, one is to try to increase the time for planning and give better decision support using *information* more efficiently and the second is to simply make the planning more easy through *standardisation*.

The information needed for planning can be grouped in several ways, in this model we have chosen to divide it by when it is used. There is administrative pre-transport information (e.g. bookings, transport instructions etc.). There is also operative in-transport information, which can be further divided into information that is gathered continuously and information that is gathered when the transport deviates from what was originally planned. Finally there is post-transport information, which is both administrative (proof of delivery, payments etc) and operative (e.g. the position of trucks and trailers, for reducing empty haulage).

No matter how good our information systems are we think however that the other way towards simplified planning is perhaps even more important. Through standardising our transport system in respect to routes and terminals we can achieve high resource utilisation with less information through simplifying the planning process and increasing the volume of goods available to fill the transport units in a more efficient way. The important thing in this, is that while we are working in this direction we will also in a way appear more flexible to the client as we dramatically increase the transport frequencies in our transport system, something that the word standardisation perhaps does not imply. In the process it will also be easier for us to provide value-adding locistics services to our clients as we focus our strength at a number of different points in Europe, wherefrom we will be able to reach almost all corners of our network in 24 to 72 hours (72 hours depending on ferries).

SCANSPED GROUP'S RESPONSE TO THE MARKET CHANGES

In response the the changing markets we are currently undertaking several interconnected projects, where the two most important are called FUTURA and CIEL. FUTURA is aimed at increasing the efficiency in the transport system through hubs-and-spokes, standardisation in the above discussion, while CIEL is an IT project mainly working with the administrative information needed to make FUTURA possible.

FUTURA

The FUTURA concept is based on:

- A new organisation in our European companies;
- A hub-and-spoke transport network;
- EDI communication internally and externally;
- A new settlement system within the Scansped Group.

The FUTURA system will use a number of transfer terminals, hubs, in different parts of Europe. Between these hubs, and thereby between most European destinations, we will have daily departures in an integrated network. The pricing will be simplified and more uniform than it is today and the timetables will be more static than today, but still provide several times higher frequencies than we currently have to many destinations. An important comment to this is that we sell a transport service to the client, promising to pick-up and deliver his goods at two specific points at certain times. In our network we can however use different routes from one time to another depending on the utilisation of transport units and terminals as long as we start and end at the agreed locations. In creating this *dynamic hub-and-spoke* we can always guarantee the service level to the client while optimising our resource utilisation.

Organisationally FUTURA is based on a customer organisation working with our clients on a local level in order to keep close to the market, and a production organisation working with the actual transport on a more centralised level in order to make the use of resources optimal. We are creating a system with small-scale market prospecting in a flow-oriented production system not using national borders as separators between companies.

In doing this we can get both the traditional vertical result units and the logistically more interesting horizontal units, focusing on the actual movement of goods and the efficiency and effectiveness of the entire network. This is important in avoiding future suboptimisations.

In working with FUTURA it is of utmost importance to have a set of information systems available and working. We have during the past few years concentrated on acquiring and developing the systems needed to implement FUTURA. As stated earlier we are relying heavily on EDI, internally as well as externally, as a means to move information between parties and systems in the network. A brief description of our information systems is given below.

CIEL and other information systems in the Scansped Group

CIEL, our Computer Integrated External Logistics system, is the main system of the Scansped Group, it is the actual forwarding system where most data on bookings, transports, clients and so forth is processed and stored. It is basically a standard system that, even though most of the functions were there when we bought it, has been further developed to suit our needs. The CIEL system is used throughout the Scansped Group in the same way, making the integration between companies simple and ensuring that goods and clients are treated the same way wherever they appear in the network.

One of the most important aspects in choosing and developing CIEL is the possibility to connect the system to other systems. As we want to see the entire information flow as a whole it is very important that the different systems can interact and supply data to each other. Therefore CIEL has been given a new EDI module, accepting all the transport related EDIFACT messages and several more specific messages from clients or different parts of the company, to and from banks, suppliers, authorities and so on. We have also naturally created links between CIEL and our Result Measurement System (RMS), a newly developed system for settlement calculations and horizontal result measurement, and can without any problems transfer data from CIEL to our planning systems, e.g. Superspin, an American network optimisation application.

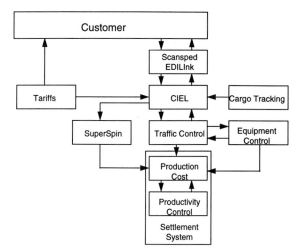

Figure 2. Interaction of information systems

Results of the changes

As we can see there is a very strong link between the different parts of the tremendous change process at the Scansped Group. Especially we need to understand the importance of the relation between physical flows and information flows to create the high performance transport system needed to stay competitive.

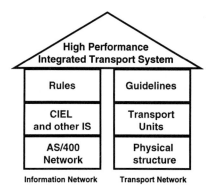

Figure 3. Basic components of Scansped Group's High Performance Integrated Transport System

We have now been working for more than two years with the design and implementation of the various physical and information systems and we are in the middle of the change process, which we have partly described in this paper (there are of course other aspects of development that could have been discussed but are outside the scope of this paper, e.g. new value adding logistics services for our corporate clients).

We are heading in the right direction and the change to FUTURA is a major part of our European development. FUTURA has now been implemented between Sweden and the UK, Sweden and Holland, Sweden and Belgium, Sweden and Switzerland and between Denmark and Norway. The results so far are good, as we have increased our service by offering daily transport services and we are using our transport capacity more efficiently. The implementation will continue. FUTURA requires volume and the positive effects of FUTURA will be even more obvious when a larger part of our traffic is operated in the same way.

17

CHANNEL CONCEPT IN A CONSOLIDATION NETWORK

Ola Hultkrantz
Bo Ireståhl
Kenth Lumsden

ABSTRACT

To achieve a high degree of coverage over large areas together with a high utilisation of resources the transportation companies have been forced to consolidate goods in larger and fewer warehouses. The requirement for a high utilisation of resources and a high capacity of the flows in the network means that the speed of the flow must be constant throughout the whole system. It is possible to achieve this by connecting a number of well defined spokes and hubs in a network to establish transport channels: the channel concept. Instead of viewing the forwarder as a supplier of transport, goods handling, warehouse space, administration, IT-systems and other activities, he should be considered as a supplier of various flows. The possibilities of taking advantage of the benefits of a channel concept in a consolidation network system have been analysed. It means a better customer service level in terms of higher frequencies and the maintenance of short transportation times. Operations should be cost-effective due to a high utilisation of the transportation equipment and avoidance of terminal handling.

A flow where cargo, information, resources and money are transferred between two geographical points, can be defined as a traffic line or a business line, the channel. It can be further divided into production lines because of network reasons such as terminal locations. A transport company with several points of dispatch and receipt, has a great number of business lines. When the cargo needs to be un- or reloaded along the route e.g. at a terminal, a production line can be defined. A business line can be made up of several production lines.

In transport companies using the channel concept within the network together with many subsidiaries and offices, there will be a need for co-ordination which often is created with a matrix organisation. The channel concept is well suited to international organisations, aiming at optimising both business operations and production performances.

Keywords: Channel concept, line based road traffic, hub and spoke network

BACKGROUND

To achieve a high degree of coverage over large areas together with a high utilisation of resources the transportation companies have been forced by technical and financial factors to consolidate goods in larger and fewer warehouses. In this way they can reduce stock expenses, lead-times and the handling cost of parcels by using modern equipment and automated facilities. On the other hand, the node system (hub) requires consolidation of flows on lines (spokes) with various systems: LTL, less-than-truck-load, and/or fixed schedule (OECD, 1992). To achieve a high degree of utilisation, the transport company collects goods from several consignors in one hub (source). The goods are then transported, through one or more relations or spokes, to the hub (sink) which has the demand.

The requirement for a high utilisation of resources and a high capacity of the flows in a network mean that the speed of the flow must be constant throughout the whole system. This is possible to achieve by connecting a number of well defined spokes and hubs in a network to establish a transport channel, the channel concept (Lumsden, 1995).

The line-based traffic system is however not very efficient in terms of resource utilisation (Buskhe, 1993). On the other hand, the hub and spoke system creates extra transport work since the goods do not go the shortest way. They go via a terminal to be reloaded with goods from other regions with the same destination. This means that the goods have to be transported a longer distance i.e. extra transport work has to be done and the transport cost increases. Some negative effects have been pointed out (Barker and Sharon, 1981). Specifically, they state that the use of transfer terminals means that the freight must be physically handled at break-bulk which is costly. They further point out that handling takes time, averaging a day of service loss in a break-bulk station.

In the manufacturing industry a number of new concepts have grown such as JIT, TQM, BPR etc. which have created a focus on the material flow through the company. The importance of logistics in the industry, the increasing function it plays within every firm, and the place it now occupies in their structure, demonstrates clearly that control of flows has become strategic. It has been pointed out concerning JIT that "...the plant no longer functions as a step-by-step process that begins at the receiving dock and ends when finished goods move into the shipping room. Instead, the plant must be redesigned from the end backwards as an integrated flow" (Drucker, 1990).

The transportation industry is confronting a major change because of deregulation and increasing global competition. This implies that old ideas and ways of thinking need to be replaced by new ones (Ericsson et al., 1995). One concept in line with these requirements is to include a channel concept in a consolidation system network.

It is important to understand the basics of LTL operations to appreciate the complexity of the problem. LTL networks are characterised by a set of terminals that handle consolidation. We treat the beginning of the journey of an LTL shipment as starting at an end-of-line terminal. However, the shipment begins at the consignor where the goods are collected by a citytruck from the local office and transported to an end-of-line terminal. Here the process of moving the goods through the network begins. At the end-of-line terminal the goods might be unloaded and reloaded onto another truck that will take the goods into the network. This truck will not necessarily take the goods to the hub close to the consignee. The goods could be transferred to a truck at a terminal on its way to the consignee's end-of-line terminal.

A LTL network uses a set of transfer terminals to serve as collection points for freight originating and terminating at terminals in those regions (Lumsden,1995). The satellites around a transfer terminal form a hub and spoke network which performs the primary consolidation function, feeding high density lanes that join each pair of breakbulks. Even larger transportation systems could be discussed in this way.

From the perspective of network planning, shipments are treated as originating and terminating at the end-of-lines. Trailers always move directly between end-of-line and transfer terminals. This pattern of moving from end-of-line to transfer to end-of-line terminal(s) is often referred to as the standard routing for LTL carriers (Braklow et al., 1992). In many cases it is possible to use a more efficient transport route. Assume that there are enough goods at the end-of-line terminal near the consignor so they can fill a trailer going directly to the destination end-of-line terminal without passing the ordinary transfer terminal. This leads to less handling at the transfer terminal. Eliminating the need to handle a trailer not only saves in cost, but it can also reduce transportation time.

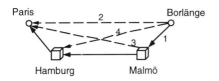

Figure 1. Four different transport alternatives that can appear in a normal hub and spoke system. (1) The normal route is Borlänge-Malmö-Hamburg-Paris. (2) There are enough goods at Borlänge to fill a trailer going directly to Paris. (3) The transfer terminal at Malmö has got enough goods to fill a trailer for direct transport to Paris. (4) The end-of line terminal in Borlänge sends the goods direct to the transfer terminal in Hamburg where they are consolidated with other goods and transported to Paris.

Four problems regarding the routing alternatives are to be discussed (Braklow et al., 1992). First, the cost of handling freight at Hamburg may be expensive for several reasons, e.g. different salary rates, implying that the savings in handling at Malmö may be more than offset by higher handling costs at Hamburg. Second, Hamburg may be close to capacity, creating additional congestion problems. Third, it is possible that the change in routing actually increases the overall distance a trailer must cover. Fourth, a systematic problem with directs run out of end-of-lines. As a rule, the load average per trailer for direct trailers departing from end-of-line terminals is lower than for direct trailers departing from transfer terminals. Transfer terminals have both a greater volume and a broader mix (small and large, light and heavy) of freight that makes it possible to pack more kilos into a trailer.

AIM AND SCOPE

The aim of the study is to create an organisational model for applying the channel concept in a consolidation network system and to analyse the possibilities of taking advantage of such a model in a working condition. This should mean a better customer service level in terms of higher frequencies and maintaining short transportation times. Operations should be cost-effective due to a high utilisation of the transportation equipment and avoidance of terminal handling (Buskhe, 1993).

METHOD

This paper is based on case study methodology. This is determined mainly by the character of the research problem, control of the events and type of phenomena studied (Yin, 1994). A model of the concept was created out of existing demands and of current transport systems. This was then tested in a new system based on this model. A number of channels were studied, all of which use a new way of organising the transportation. The empirical data is a result of the implementation of the model.

THE TRANSPORT NETWORK

The flow system

The assignment of the forwarder is to provide the service of transport from one point to another and to move almost anything the customer wants to be moved between these two points or other designated points. Moving goods also means that the forwarder can handle the goods and will give it added value through the performed services. One such service is administration and financing by means of standardised information transfer and network systems, such as databases and EDI.

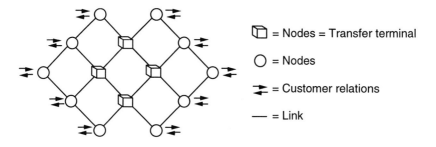

= Nodes = Transfer terminal

= Nodes

= Customer relations

— = Link

Figure 2. The flow in a network

Therefore, instead of viewing the forwarder as a supplier of transport, goods handling, warehouse space, administration, IT-systems and other activities, he should be considered as a supplier of various flows.

The channel concept

Looking at the assignment of the forwarder this way will involve viewing the work in terms of flows and the way to handle these flows. A flow will emerge or be produced whenever there is a traffic line or a transport facility between two or more designated points. Once a traffic line is established it could be used as a kind of tube or a pipeline for the flow of goods and the flow of services. This includes the flow of administrative data that you want to see moving through the traffic system.

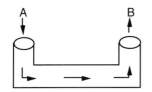

Figure 3. A single flow

Depending on the customer location, flows can emerge in several places simultaneously. Flows will cross each other and form networks. In such situations there will be a constant need to find the best traffic line for the transport service offered by the forwarder with regard to the particular customer. A traffic line is characterised by scheduled and frequent departures. A flow concept incorporating the channel concept is then identified (Figure 3).

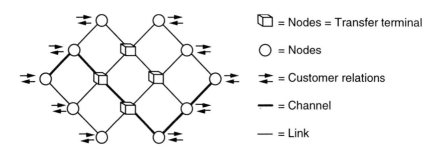

Figure 4. The channel concept (A to B) in a flow oriented consolidation network system

For each flow, a transport company has to consider what price it should charge the customer, what loading factor the flow of units should and will have and the utilisation rate of the equipment. A transport company should also consider whether there should be any terminals along the flow line. Here, it is important to have a good balance of the flow of freight in all relations, i.e. to have as few empty kilometres as possible.

Flows of cargo, information, resources and capital

Flows are normally identified as transmitters of cargo. This transmission can be performed by means of load carriers, terminals etc. In this way, a traffic system is created. The flow could also in a broader sense act as a pipeline for various kinds of information, such as market data and statistics. Financial transactions that arise from a cargo movement could also be considered in terms of flows. Consequently, an information flow system (computer systems and communication networks) as well as a financial flow system (tariffs, settlements, payments, profits and losses) will be within the flow system. These systems move parallel with the cargo to which they are related. Resources consist of resources that are connected to the beginning or the end of the activities or load-carrier etc. The resource system then consists of two parts, one that does not move at all and one that circulates within the network.

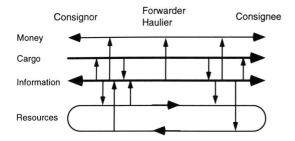

Figure 5. The four flows in the transportation chain (Kanflo and Lumsden, 1991)

THE CHANNEL CONCEPT IN A CONSOLIDATION NETWORK

Business lines and production lines

A flow or a traffic line, where cargo, information, resources and money are transferred between two geographical points, can be defined as a business line. It can be further divided into production lines because of network factors such as terminal locations. A transport company that is constantly working in a flow network, with several points of dispatch and receipt, has several business lines.

When the cargo needs to be un- or reloaded along the route (between A and B) at for instance a terminal, we can define a production line. A business line can be made up of several production lines.

The flows could also be seen from a dispatching/receiving point of view. A transport company dispatches cargo every day. The appropriate information and financial systems are attached to these activities. The dispatching/receiving orders are executed through different business lines.

On behalf of its customers, the transport company must take two consignments from dispatching offices to receiving offices (C and D respectively). To do this it makes use of its production lines and tries to organise the flows in co-operation with its own offices or partners (B and D).

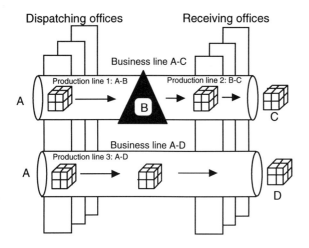

Figure 6. Definitions of business and production lines (The business lines are A-C and A-D. The production lines are A-B (because of the terminal), B-C and A-D. The A-D production line in this example is identical to the business line)

The concept requirements of a flow-oriented organisation

The focus has so far been on the need for a different view of the transport assignment in terms of flows. A perspective of flows of cargo, information, resources and money has been introduced. This is the basis for the traffic system (load carriers, terminals etc.), the information system (computer systems and communication networks), and the financial system (tariffs, settlements, payments, profit and losses).

The importance of organising and supervising production lines together with their flows and related systems has been indicated. It has been pointed out that every transport office must keep track of the local business it creates through its dispatching/receiving offices.

It is in relation to this business requirement that there must be an emphasis placed on the need to *create results*. This can be divided:

- horizontally - through the business lines and production lines;
- vertically - through the orders from customers to the national company's local office.

To organise a flow-oriented business concept there is a need for a firm management capable of working according to both the vertical and horizontal structure. The need for sales administrators in charge of receiving and dispatching order volumes is evident. In other words people with customer responsibility who generate income for the company should be predominant.

There will also be a need for a specific person responsible for the planning and production of the transport, i.e. the traffic system itself, including the physical cargo movement from A to B. This means handling the cargo, securing the appropriate vehicles and load carriers, organising terminals if necessary, and deciding on the most cost-efficient production line for each particular customer order.

The responsibility for the fulfilment of the task and the improvement of profits (customer revenues minus production costs) on each of the business lines must be dealt with. The person responsible for this flow will have to decide on the tariff levels, suggest alternative business lines and organise and execute market activities.

The matrix organisation

In large transport companies with a large flow in the network and many subsidiaries and offices, there will be a need for co-ordination and a general survey of operations. This is often presented in terms of a matrix organisation (Irestähl, 1996).

A matrix organisation has two important messages:

- It states that the day-to-day responsibility for the local-national-company, its personnel and its vertical results, remains with the Managing Director and his local company: production responsibility. The national organisation has full responsibility for profits within its defined market place. In this respect, the matrix organisation does not differ from a conventional organisation.

- It states that there will be a need for network functions, in this case called specialists, whose tasks are to develop and co-ordinate assigned functions. These specialists have strategic responsibilities as they represent group values and policies that must be followed by all company units, regardless of their size and location. The group management has assigned certain overall responsibilities to these specialists. For the benefit of the group, the specialist may intervene in strategic matters such as co-ordination of business lines or/and products. Besides their strategic role, which is mandatory, the specialists also have a responsibility for the budget for the horizontal results: business responsibility. The functions are not geographically bound, as those of the national organisation, but involve several company units and their market places.

The matrix works both vertically and horizontally. The concept is well suited to international organisations, aiming to optimise both group operations (horizontal results) and national performances (vertical results).

The relationships between the vertical and horizontal responsibilities can be characterised graphically (Figure 7). An attempt to explain this includes the relationship between national market place organisations and group specialists. The co-ordinating function in this example is that of a business line comprising marketing and production, responsible for a standard transport company, that works horizontally with several subsidiaries or local units belonging to the group of companies.

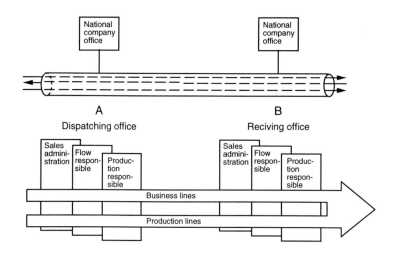

Figure 7. The matrix organisation

Flow-oriented transport organisations

A group of international companies, in the transport business, work through a network of subsidiary national companies. The subsidiaries normally report to a head office. The purpose of the head office is to manage and assess the organisation in accordance with overall company goals and general policies. Attached to the head office are staff functions with the aim of securing the assigned overall tasks of the group. These functions could be Group Finance, Group Traffic Flows, Group IT, Corporate Personnel, Group Marketing etc.

Each national company has a direct relationship to its customers in the local market place. A market place is normally defined as a country. The customer should be well integrated in the traffic flow system of the transport company. The traffic flow system is represented by many business lines offered to the customer by the local transport company (Figure 7). Business lines exist between any dispatching and receiving office. In the transport company the business lines are made up of several production lines. There must be overall horizontal responsibility for the smooth functioning of the business lines system, i.e. the overall production and marketing system of the transportation company.

CASE STUDY - FUTURA AT SCANSPED

Scansped is a member of Bilspedition Transport and Logistics (BTL). BTL is the biggest forwarder in Scandinavia and one of the five biggest in Europe. Scansped is established in Europe with its own offices in almost 70 cities. In the traffic system they also co-operate with a number of local partners. The Scansped network before Futura was characterised by independent traffic lines. At present Scansped has around 5,000 scheduled flows or traffic lines throughout Europe.

The implementation of the new way of organising their work has started with the traffic between the Scandinavian countries and the UK, Belgium, Netherlands and Switzerland. At the moment, Scansped has about 180 business lines, 90 in each direction. 48 terminals are involved in the Futura concept and during 1996 all of Scansped's terminals are supposed to work according to the Futura concept. The goal is that every business line shall have daily departures (weekdays) and at the moment, all of the business lines that are in use, have achieved this goal.

The volumes between the most important terminals in the Futura system can vary from 32,000 tons to 2,000 tons, with an average of 11,400 tons. The loading factor is on average 81.5%, minimum 27% and maximum 119%, calculated for 24 ton trailers, with an average difference of 30.4% for a round-trip, minimum 7% and maximum 83% (Table 1).

Table 1. Volumes between 10 important terminals in Scansped's network. Note that the volumes for each terminal consist of more than one business line and that the loading factor is calculated for a 24 ton trailer.

Traffic between terminals (A, B, C, .., J)		Number of consignments	Weight (10^3 tons)	Loading factor (%)	Difference (%)
Dispatching	**Receiving**				
A	B	4146	10	95	
B	A	9062	16,1	102	7
A	C	3505	7,5	95	
C	A	4956	5,8	105	10
A	D	2044	6,6	76	
D	A	1992	5,4	52	24
E	D	5528	23,4	85	
D	E	5924	11	64	21
A	G	1944	5,4	73	
G	A	3084	12,7	90	17
E	G	6636	32	86	
G	E	6072	14,3	49	37
A	F	232	2	71	
F	A	2548	7,3	98	27
E	F	4512	14,4	96	
F	E	6608	20,8	119	23
H	I	5092	13,6	96	
I	H	952	3,4	41	55
J	I	3124	13,3	110	
I	J	1664	3,4	27	83
	Average	3198	11,4	81,5%	30,4%

In Table 2 the data is arranged according to the difference in loading factor between dispatched and received goods. From this table it can be seen that there is a clear trend. The terminals with a higher number of consignments have a better matching between dispatched and received goods. This implies that it is more efficient to balance the traffic if a number of transfer terminals (hubs) are used to increase the quantity in the used traffic lines. This needs to be investigated further.

To standardise the way of working, the company has tried to use a highly-developed standardisation. In the Scansped world, for example, the traffic system is expressed through the standardisation of its production lines, the information system through CIEL and EDI, and the financial system through uniform payment and settlement in their RMS (Result Measurement System) (Ireståhl, 1996).

Table 2. Volumes between the 10 terminals aggregated on dispatched and received goods for each terminal. Note that the volumes for each terminal consist of more than one business line and that the loading factor is calculated for a 24 ton trailer.

Office (terminal)	Dispatched			Received			Difference		
	consignments	Weight (10^3 tons)	Loading factor	consignments	Weight (10^3 tons)	Loading factor	consignments	Weight (10^3 tons)	Loading factor
J	3124	13,3	110%	1664	3,4	27%	1460	9,9	83%
H	5092	13,6	96%	952	3,4	41%	4140	10,2	55%
F	9156	28,1	108%	4744	16,4	84%	4412	11,7	24%
E	16676	69,8	89%	18604	46,1	77%	-1928	23,7	12%
C	4956	5,8	105%	3505	7,5	95%	1451	-1,7	10%
B	9062	16,1	102%	4146	10	95%	4916	6,1	7%
A	11871	31,5	85%	21642	47,3	89%	-9771	-15,8	-4%
G	9156	27	70%	8580	37,4	80%	576	-10,4	-10%
D	7916	16,4	58%	7572	30	80%	344	-13,6	-22%
I	2616	6,8	34%	8216	26,9	103%	-5600	-20,1	-69%
Ave.	7962	11,4	86%	7962	11,4	77%	3460	12,3	29,6%

CONCLUSIONS

Today it is very common in the transport industry (airlines, maritime, trailer etc.) to use a hub and spoke route structure for taking the advantage of a consolidation system and to have more frequent departures. In the manufacturing industry, a number of new concepts have emerged such as JIT, TQM, BPR etc. which have created a focus on the material flow through the company. The emergence of logistics in the industry, the increasing role it plays within every firm and the place it now occupies in their structure demonstrates clearly that control of flows has become strategic.

By combining the flow concept from the manufacturing industry - and in this way integrating the distribution in the flow - with a distribution network such as a hub and spoke network, a number of advantages can be attained. The planning at the involved terminals becomes more efficient when you know which way the goods are shipped and when they are permitted to take short routes in the distribution system. The consolidation of goods at the transfer terminals makes it possible to get more goods on each traffic line (spoke) which in turn makes it possible to achieve a higher loading factor.

When the responsibility is distributed along the flow it is possible to decide the origin of each goods unit, e.g. the goods can be defined concerning source hub as well as the route the goods are transported through the network. This means that the responsibilities for problems which could occur can be specified.

By using distribution channels with fixed departures and fixed capacities the planning, both for the transport company and its customers, will be simplified. The result of this new concept will be that the demand for frequent departures, short delivery times and at the same time a high utilisation of resources can be managed within an existing consolidation network.

REFERENCES

Barker, H. H. and Sharon, S., *From Freight Flows and Cost Patterns to Greater Profitability*, Interfaces, Vol 1, pp. 4-20, 1981

Braklow, J. W. et al, *Interactive Optimization Improves Services and Performance for Yellow Freight System*, Interfaces, 22:1, Januari-February, pp. 147-172, 1992

Buskhe, H., Resource utilization in line-based road traffic, Department of Transportation and Logistics, Chalmers University of Technology, Göteborg, Sweden, Rapport 23, 1993 (in Swedish)

Drucker, P., *The emerging theory of manufacturing*, Harvard Business Review, 68: 97-102

Ericsson, D. et al, *Virtual Integration - Information technology the enabler in globalization*, Unisource, 1995

Ireståhl, B., *Scansped Reports, Futura*, Gothenburg, Sweden, 1996

Kanflo, T. and Lumsden K. R., Informationflow in the transportation chain, Department of Transportation and Logistics, Chalmers University of Technology, Göteborg, Sweden, 1991 (in Swedish)

Lumsden K. R., Transporteconomy - Logistics models for flows of resources, Studentlitteratur, Lund Sweden, 1995 (in Swedish)

OECD, *Advanced Logistics and Road Freight Transport*, Road Transport Research, OECD, Paris, 1992

Taha, T. T., and Taylor, G. D., An integrated Modeling Framework for Evaluating Hub-and-Spoke Networks in Truckload Trucking, Logistics and Transportation Review, Vol 30, No 2, pp. 141-166, 1994

Yin, R. K., *Case study research*, Sage, Thousand Oaks, USA, 1994

18

TRACKING AND TRACING AT AN INTERNATIONAL HAULIER: WHERE ARE WE NOW, AND WHERE DO OUR CUSTOMERS WANT US TO GO ?

Geert Hoek
Kjell Högberg

ABSTRACT

The goal of tracking and tracing is to achieve more process control over goods during transport. In international transport, the express services successfully introduced this logistic tool. Tracking and tracing is also a marketing tool in order to distinguish oneself in a market where structural overcapacity results in lower profit margins. Offering more quality through the use of a high standard process tool can well be the Unique Selling Point in the next few years. The everyday practice of applying an example of a tracing system, its benefits and drawbacks, is described and secondly a draft will be presented of the customer's wishes on the subject of T&T for the Scansped Group. This group, whose main activity is European truck groupage, is a division of Bilspedition Transport & Logistics (BTL).

Keywords: Tracking and tracing, BTL, groupage

INTRODUCTION

Lately, big international shippers started demanding from their carriers a similar control over their flow of goods, as they themselves have within their plant. There is definitely a demand for a system that will enhance control over goods during transport. One of the tools to achieve better process control, could well be the application of a Tracking and Tracing system.

In practice, and also in literature, a Tracking and Tracing system is often simply referred to as a system that enables a carrier to locate goods during transport. However, a Tracking and Tracing system is more than that. It should be used as a process control tool, it should be able to monitor and control the goods flow.

The Scansped Group is one of the divisions of the Bilspedition Transport and Logistics group (BTL). The Bilspedition Group is not only Scandinavia's largest transport and logistics company, but also one of Europe's. The majority of the business activities is concentrated in the EU, Norway, Central and Eastern Europe. Including associated companies, Bilspedition has offices at 520 locations in 31 countries. All means of transport are used in transport production - lorries, railways, ships and aircraft. The Group's customers are industrial and trading companies with predominantly high value goods such as consumer goods, engineering products and components. Approximately 18 million tonnes of goods are transported annually in the Group's networks.

The Scansped group's core activity is transporting groupage goods throughout Europe. Groupage is the receival of goods from several customers in a terminal, consolidating them and transporting them to a destination. The Scansped group is using the so called hub and spoke structure. This means that most consignments are collected by domestic distribution, brought to a terminal, transported by an international truck to a foreign terminal, and delivered by domestic distribution.

The current tracing system of a Scansped terminal in Holland is analysed. Similar systems are applied in other Scansped terminals. Secondly, a description is given of the customers requirements on T&T, based upon a survey within the whole Scansped group.

WHERE ARE WE NOW ?

Description of the system

A Scansped terminal does already apply a tracing system for a couple of customers. Figure 1 shows the seven steps of the system. It is not integrated in the existing information system and requires a considerable amount of man capacity without even entirely satisfying the customer. External parties (the consignee or shipper) do still notice exceptions (e.g. delays) earlier than Scansped. This system does not rely on complex telematics. A description of the procedure can be found below. Most blocks are activities by the Dutch Scansped terminal, while boxes 4 and 5 represent activities by the other terminal involved. Following the seven steps will lead to a feedback of the actual delivery date of a consignment to the shipper.

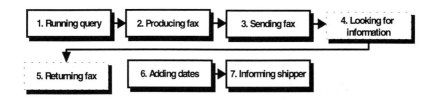

Figure 1. The seven steps in the current system

Step 1 Running query

As all consignments are put into the information system, a daily query is run in this system in order to select all consignments of a particular shipper, requiring status information.

Step 2 Producing fax

The results of the query are transferred to a spreadsheet, where consignments are selected on agent/terminal. The used agent/terminal depends on the volume of the consignment and the destination. Which agent or terminal has been used can be retrieved from the master number (the number of the trip) of the international transport, which is linked to a consignment number. A sheet is made for every agent/terminal, on which the consignments are described.

Step 3 Sending fax

The sheet is faxed to the last Scansped terminal or agent in the network for that consignment.

Step 4 Looking for information

The other terminal has to look for the information and writes this information on the fax.

Step 5 Returning fax

This fax is to be returned on the day of delivery or the next day before 11.00 am. The actual delivery date and an eventual reason of non-conformity are written on the form and sent back.

Step 6 Adding dates

All faxes are collected and the actual delivery date is added in the spreadsheet. The consignments with a return answer are transferred from the list of consignments waiting for a reply (the agents file) to the list of consignments ready for feedback to shipper (the shipper file).

Step 7 Informing shipper

The shipper file is printed and faxed to the shipper. The shipper receives an overview of his consignments on a daily basis.

Using this system results in an elementary feedback to a few customers. In Figure 2, an overview of the system is shown.

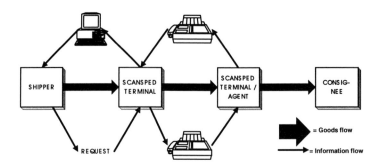

Figure 2. Overview tracing method

Analysis of the current system

The seven steps from the previous section demand effort and co-operation from all employees, not only employees from Scansped itself, but also from other terminals and agents involved. The benefit of this system is obvious: supplying information for shippers and measuring service levels. Despite the efforts, this procedure led to the following problem areas:

Time consuming

These seven steps are time consuming. Working in the spreadsheet, moving data from one file to another, checking the completeness of the consignments, faxing, processing the answers and moving data in the spreadsheet file result in a considerable work load.

Limited measurement

- The delivery status is measured the day after delivery or later. There is no possibility for rescheduling.
- There is no guarantee that there is actually a control of the goods flow.

Limited capacity

This service cannot be extended to many more customers. Using the spreadsheet in the personal computer is now already causing process delays and printing problems.

Lack of co-operation

Six weeks after starting this process, one agent refused to co-operate any longer. This agent ships a very high percentage of the consignments to Germany. The co-operation of some other agents was gained only after many phone calls. This results in a large time span between sending a fax and receiving an answer. Return faxes from Italy for example, do often have a delay of two weeks or more.

Pro-activeness

Some agents/terminals return their fax before the actual delivery. The date filled in represents their estimate of the actual delivery date. This pro-activeness results in an advantage and a disadvantage. The benefit is the possibility to inform the customer in case of a difference between the actual delivery date (which is the agents/terminal estimate) and the date that was estimated and promised while booking the job. Informing the shipper is called negative reporting. The drawback of filling in this date before the actual delivery is obvious: the estimate may be wrong and the shipper will be provided with incorrect information.

Little information on non-conformities

The last column, in which one is supposed to fill in an eventual non-conformity, can be neglected in most cases. This means that there is no feedback on the condition of the goods; a consignment may be delivered on the right date, but non-conformities such as damage, shortages and overloads should also be reported.

Input errors

Using the information from the information system requires an accurate input from the employees who handle the input of consignment information into the system (job booking). This accurate input is needed for running the query and for communicating with the agents/terminals. Examples are the input of wrong client Ids or a wrong estimate of the delivery date.

Follow-up

After using this system for a reasonable time span, the collected data were used for input in a performance analysis. By comparing the estimated and the actual delivery date, the service level of every agent/terminal is measured for every shipper. However, it is acknowledged that the problems mentioned do have an impact on the performance analysis. Due to these reasons, only 37 % of the consignments to be traced, are actually traced.

Comparing the promised delivery date and the actual delivery date is not an answer to the question why deviations exist. As long as there is no information on the exact description and location of a non-conformity, solving the cause of a non-conformity is impossible. The non-conformity could have occurred during collection, during national transport, at the Dutch terminal, during international transport, at the agent or foreign terminal, during national transport or at the delivery address.

Costs of the current system

The described procedure is rather time consuming and seems rather expensive. Therefore, the costs of using this system was estimated. It is stressed that these costs are related to the Dutch terminal only. Activities in other terminals, such as sending the return faxes, were not taken into account in this calculation. Based on six weeks of measurements, the costs of tracing a consignment, using the described method, is about SEK 22 for every consignment actually traced.

This average of SEK 22 is composed of the following two main components:

Labour costs

A practical consequence of applying this system, is the fact that it is time consuming. Based upon a six weeks measurement the activities shown in Figure 3, are distinguished. The percentages in Figure 3 are percentages of the total labour time spent using the current tracing system in the Dutch terminal.

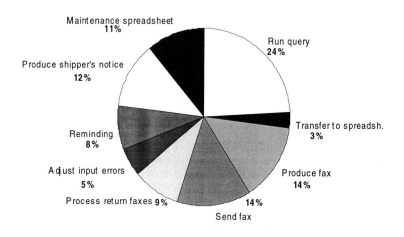

Figure 3. Division labour time of current tracing system in the Dutch terminal

On average, these activities take about 145 minutes a day. The hour rate of the employee is around SEK 160. This results in an average labour cost per day of SEK 388.

Communication costs

The second component of the costs are the communication costs. Everyday, around nine faxes and five phone calls are made. Faxes have an average duration of around fifty seconds, phone calls have an average duration of four minutes. The average cost of communication is around SEK 4.80 per minute. This results in average communication costs of SEK 104 per day.

Based on an average of 22 actually traced consignments per day, and total tracing costs of around SEK 492, one can conclude that the cost of tracing a consignment for the Dutch terminal is around SEK 22.

During the analysis of the current system, imperfections of the current system were already discussed. A goal of Scansped is to build another kind of T&T-system which does not have these imperfections, and which is the closest approach to the customers needs. Before such a system can be built, the need for such a system from customers and the Scansped branches needed to be investigated. By means of a questionnaire among Scansped organisations throughout Europe, the need for such a system was mapped out. This need can be divided into two different needs: the internal and the external need. The internal need is the need from the Scansped organisation itself, and the external need is the demand for a T&T-system by customers. In this article, we will focus on the external need.

WHERE DO OUR CUSTOMERS WANT US TO GO ?

In a survey (Hoek, 1996) managing directors, quality managers and branch managers of Scansped organisations were asked to fill in the questionnaire. 54 Forms in total were distributed. Of these 54 forms, 38 were returned, meaning a response rate of 72%.

After registration, every response was categorised on country of origin. For every country, averages of the responses were calculated. In order to find an average for the whole of Europe, a European average was calculated based upon the number of consignments in every country. If for example Scansped Sweden transports ten times the amount of Scansped Poland, the impact of the Swedish answer on the European average needs to be 10 times that of the Polish answer.

The demand can be separated into different aspects. These aspects are described below.

Availability of T&T

The demand of T&T will be influenced by what competitors (other suppliers) offer. Offering T&T in a country will probably lead to an increase of T&T-demand in that particular country. That is why one of the questions in the survey referred to the number of competitors working or developing any kind of Tracking and Tracing systems.

On average, T&T is supplied in every Western and Northern European country. Especially in Holland, Germany and the UK, quite some companies are using T&T already. Only in Eastern Europe, T&T is not used yet. If separated into the three following groups: express services, groupage and full loads, most companies who are using any kind of T&T, are express companies.

Asking for the number of companies, which are currently developing T&T, led to the conclusion that most companies which are now developing T&T, are groupage companies. As European truck groupage is one of the core activities of Scansped, this result should be seen as another encouragement for developing a T&T-system.

Quantity of demand

As most Scansped branches do not use T&T-system yet, one way of measuring the information demand is by estimating the relative number of consignments, for which a Proof of Delivery (PoD) is being asked. Of course this way of measuring excludes, for example, the number of phonecalls made to inform about the status of a consignment. On average, external parties (shippers / principals) demand a PoD for about 6 % of the total number of consignments. This means that a PoD has to be given for about 90,000 consignments every year.

Another way of measuring the degree of customer demand, is to measure the percentage of customers who ever asked for a T&T-system, and the number of consignments that these customers represent. Comparing these two figures also indicates the average size of the customer. About 8% of the customers, who ever asked for a T&T-system, represent about 20% of all consignments. This means that mainly bigger customers are interested in the use of a Tracking and Tracing system.

Detail level of demand

In the current situation, information demand is mainly restricted to the information given on PoDs. For a small group of customers however, a so called negative reporting is being used. This means, that in case of exceptions, the shipper or consignee is informed about this deviation in lead time. In some countries, such as Russia, a customer is always informed beforehand about a delivery.

The necessary detail level for a T&T-system, is based on the following classification :

a) Proof of delivery. In this T&T-system, the only registered status is that of delivery (Tausz, 1994).

b) Registration of number of physical movements. For example in case the consignment moves from the terminal into the international truck, the event is registered (Watson, 1995).

c) Continuous monitoring. The status of the goods is monitored on-line, for example by satellite monitoring (Tensor Technology, 1995).

Figure 4 shows that choice b is the most frequent answer. Registration of a number of physical movements is the most frequent mentioned solution.

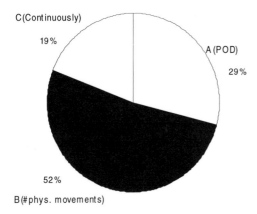

Figure 4. The percentages of responses per satisfactory detail level of T&T

Trend in quantity of demand

Another issue is the trend in customer demand. Is this demand for consignment status information increasing, unchanged or decreasing? The demand is rapidly increasing in Holland, Belgium, and Sweden. Unchanged or only slightly increasing is the demand for customer information in Poland, Czechia, the United Kingdom and Russia. Most mentioned was the category 'slightly increasing' (49% of the respondents), closely followed by 'rapidly increasing' (45%). The remaining 5% of the respondents answered that the demand for consignment information did not change at all.

System access

The demand of customers to access the system themselves in order to check the status of their consignments was measured. The estimate is that about 10% of the customers might want to use in future this way of access. One out of every four respondents however, answered that none of their customers was interested in such a customer access. Customer access could thus be a part of a T&T-system to be introduced, but will probably be used by a small number of clients.

Customer type

In order to find out which type of customer is most interested in consignment status information, customers were divided into three groups:

a) Key accounts;

b) Regular clients;

c) Incidentals.

The far majority of the respondents answered that key accounts are most demanding, followed by the regular clients and finally the incidental clients. This is a confirmation of the figures, mentioned in the quantity of demand for status information, where it was stated already that the customers who ask for T&T are mainly the bigger companies.

The type of goods

A second aspect of a customer is the type of goods that are to be transported. 45% Of the respondents, representing 53% of the consignments, answered that there is a link between the value of the goods and the interest in a T&T-system. The presence of this link means that T&T is an important tool in keeping and attracting customers with high value products, such as computer hardware.

The destination of transport

Around 30 % of the respondents answered that the need for consignment status information depends on the destination of the transport. Especially destinations in Eastern Europe cause more demand. Thus, the consignment information demand is (weakly) related to the destination of transport. In the section on availability, the conclusions were that T&T-systems were only available in Northern and Western Europe, and not in Eastern Europe. This means that in countries where the demand for T&T is highest, the supply is lowest and vice versa.

Other effects

Of course, a Tracking and Tracing system has its impact on the process of goods transport. But are there other links to be identified?

Two links that were integrated in the survey are: the number of customers and the impact on tariffs. If the subject should increase as a side effect of the introduction of T&T, 100 points were given, if the subject remained unchanged 0 points, and in case of a decrease -100 points. This led to the following scores:

The number of customers: 64

The impact on tariffs: 18

Introduction of T&T will have a positive impact on the amount of customers, although there is only a small chance that extra efforts for T&T will result in higher tariffs to be charged.

CONCLUSION

Where are we now ?

The currently used tracing system at some Scansped terminals is capable of partially satisfying the needs of a limited number of customers. The costs are about SEK 22 for every traced consignment. Some drawbacks of the system, such as the unwillingness to cooperate, reveal that introduction of a more systemised system of T&T will be a huge task and will not eliminate all problem areas.

Where do our customers want us to go ?

There is definitely a need for a T&T-system. The requirements mentioned are mainly from above average sized customers. The detail level of the information demand is best met by a system which will work based on a registration of every physical movement. For a normal groupage consignment, this means that six movements should be measured. System access is only necessary for a relatively low percentage of the total number of customers. Offering T&T can be used as a unique selling point, but it will probably not affect price levels. Charging extra for offering T&T will not be in widespread use.

REFERENCES

Bilspedition, *Bilspedition Annual Report 1994*, Skandia Tryckeriet, Gothenburg, 1995.

Hoek, G.L.H., *Tracking & Tracing*, preparation assignment for graduation project, TU Eindhoven, December 1995 (in Dutch).

Hoek, G.L.H., *Tracking & Tracing: More than Locating*, Final Research Project TU Eindhoven, Eindhoven, 1996.

Tausz, A., *The high stakes game of keeping tabs*, Distribution, December 1994.

Tensor Technology, brochure 1995.

TNT Expres, *Transport and Distribution in the single market*, TNT Expres in association with CBI Initiative, 1992.

Watson D., *Ciel /400 Executive Overview Tracking and Tracing*, CSI Leeds, 27 February 1995, IDCQ523824/*INTDOC (confidential).

19

HARBOUR OPERATIONS IN A ROLL-ON ROLL-OFF SERVICE INDUSTRY

Carl-Johan Lonntorp

ABSTRACT

In this paper, harbour operations and the impact of advanced information systems is described for the world's leading ferry company: Stena Line.

Keywords: Stena Line, harbour, ferry, roll-on roll-off

INTRODUCTION

During the course of only a few years, Stena Line has developed from being a regional Scandinavian ferry line into the world's leading ferry company.

The Stena Line network includes 15 prominently located ferry routes in North West Europe, in the commercial areas of the Kattegat, the Skagerrak, the English Channel and the Irish Sea. In May 1995 this network has been extended further with the opening of a connection to Poland between Karlskrona and Gdynia.

The shipping fleet comprises 33 ferries, mainly combi-ferries for passengers, cars and lorries, but also purpose built freighters for cargo traffic. Each year some 15 million passengers travel on the ferries besides 2.5 million cars and almost 1 million lorries, trailers, containers and wagons.

Our business philosophy and the basis of our operation is to provide rapid, reliable sea transports, quay to quay. We run these operations using modern passenger and freight vessels.

Our market-oriented organisation gives us contact with our customers: contracting parties and booking staff. At all our destinations we have staff who will take good care of all units and make sure that our operations are characterised by a commitment to quality and service. This involves anything from having the right ship on the right route and the right list of sailings, to handling all goods types professionally.

The market for transport travel and freight is likely to increase. This new trend can be attributed to the changing conditions in competition for the international ferry industry. Investments in fixed links, such as tunnels and bridges between countries together with de-regulations of air traffic and the expansion of the rail network, will offer ferry freight customers new transportation alternatives.

In order to stay competitive the ferry industry has to adapt to these changes. Stena Line as a group has already begun that process. The revolutionary high speed crafts in the HSS-project, High-speed Sea Service, developed by Stena, is perhaps the most spectacular sign of the changes to come.

But there are also other activities in Stena Line, underlining our slightly changed strategical direction. In 1996 Stena Line introduced three new HSS ferries. These investments clearly show the shift in focus, step by step during the years to come which will take place in our fleet.

Growing demand for faster marine transportation will increase the percentage of both HSS ferries and ROPAX ferries. The increased freight volumes, coupled with flexible tonnage capacity utilisation required between peak and off-peak seasons, and between weekdays and weekends, will make freight ferries with some passenger capacity more common in the future.

But our freight business will not only invest in hardware. Improved systems, new technology, organisational development and a much closer co-operation between the freight divisions in Stena Line's individual subsidiaries will be beneficial to our freight customers and sharpen our competitive edge.

HARBOUR OPERATIONS

Ferry operations are an important factor in the transport chain for freight traffic as well as for railways. Efficient, stable and quick port handling routines are of great importance for the economy of scale for the transport industry.

When introducing the HSS ferries, which is the first high speed ship able to accommodate trucks and trailers, with a cruising speed of 40 knots compared to the traditional combi-ferries 18-20 knots cruising speed, the current port logistics needs improvements. The focus is to cut down the lead-time in the port for trailers and trucks. Today's port handling often feels like a bottleneck in the shipment of freight.

The first improvement is to collect as much information as possible before the consignment arrives at the departure port. Strategically we are providing services for our customers which enable them to make their own bookings. This is achieved in two ways, firstly by EDI booking facilities to our larger customers who have their own computer systems; and secondly by our own remote system called "Freight In A Box" for our small to medium sized customers who do not have their own computer systems. There are a number of other services that we currently provide or will provide using the same technique such as invoicing, consignment status, customer statistics, news bulletin boards and e-mail to provide an all-round better service to our customers. The Internet will be another way which we are looking into to communicate with our customers.

Due to the fact that more and more information is being collected before check-in, this process is no longer a data capture point, but instead signifies the change of status informing us that the consignment has checked in for a specific sailing. The second improvement is to streamline the check-in and quayside process. From this point onwards we can divide our traffic flows into two parts: accompanied and unaccompanied freight traffic. Efforts are being made to cut down lead-time and operational costs in the port for both types of freight traffic.

The accompanied freight traffic is the freight flow which requires least involvement from port handling procedures. Accompanied port handling involves the following:

- check-in of the consignment;

- the loading of these consignments onboard the ship;

- and the unloading of these consignments at the arrival harbour.

For pre-booked consignments we are looking at improving this area in three ways. Firstly, by providing self-service, automatic check-in lanes with automatic measurement systems and weighbridges; secondly by investigating alternatives to port check-in such as allowing well placed motorway service centres close to the port to undertake this process; and thirdly by utilising an automated lane control system for the loading of a ship. The first improvement could be improved even further by allowing the driver to ring a "tele-voice check-in system" from his/her cab. This would result in the driver answering any questions regarding his/her booking before check-in, requiring only the driver's ticket to be produced at the port check-in. With smart cards becoming increasingly more popular and affordable we feel that this will be the ticket replacement of the future, and will be used to provide quayside and onboard control. Due to the fact that smartcards have the ability that we can write on them, we can also store the number of valid journies and control these.

With the collapse of military research and development due to military cut-backs, a technique that was developed for military usage called "Transponder technology" is now being targeted at the commercial marketplaces. This technology has been used in military applications like detecting "friendly" fighters, etc. Transponder technology is based upon a small microprocessor unit which can be loaded with a smartcard(s), with signals being sent to and from the unit. Although this technique is relatively new, there are a great number of possibilities for this technology coupled together with smartcard technology. For example, if all vehicles were fitted with such technology, vehicles could be automatically checked in (with a check to see whether the customer has confirmed all the neccessary details and that he has payed for his ticket), and a ship could be automatically loaded with very little human intervention. At the port of Dover in the United Kingdom we ship two and a half times as many consignments as on any of our other routes. Because of the frequency of sailings departing from this port and the short crossing time most of the traffic is not booked. In this situation specific locations on the roads leading into the port could be used together with transponder / smart card technology to provide us with the basic details before the vehicle arrives at the port for check-in. The problem is until this technology is widely available there would be negligible payback. Thus, we feel that the key is to utilise this technology at the right time and we do not feel that this will happen for the next couple of years. It is also important that standards are quickly formed within this emerging technology so that the same equipment can be used in a standard way for many applications by different users.

The unaccompanied freight traffic requires more involvement from port handling procedures. Unaccompanied handling involves the following:

- delivery of consignments to the departure harbour and check-in;

- the storage of these consignments before shipment;

- the loading of these consignments onboard the ship;

- the unloading of these consignments at the arrival harbour;

- the storage of these consignments before collection;

- and the collection of them at the arrival harbour.

Stena Line is currently investigating the usage of bar code technology for an interim period and later, as mentioned above, utilising smartcard and transponder technology. This will provide a higher quality service with no outshipments and complete check-in, loading, unloading, and checkout times to be able to reduce the port lead times even further. A length measuring system has already been installed. Additionally, an automatic trailer/container video check system is in the pipeline to provide more control over damage claims and a reduction of personnel required for this activity. This system is already being used by Harwich International Ports in the United Kingdom who deal with our container handling to and from Zeebrugge in Belgium.

For both accompanied and unaccompanied traffic there is a greater tendency to send information to statutory authorities electronically by EDI or other means. One example of this is a system called OASIS run by HM Customs in the United Kingdom where information is electronically sent to their database for all consignments being shipped to or from the UK.

In conjunction to this we are also installing a common freight system within the whole Stena Line Group network which will give future benefits to our customers such as being able to book on any route from all locations, multi-leg booking facility, one customer number for all routes, being able to receive one invoice showing consignments shipped on all routes, etc.

At Stena Line we work in a business process oriented way where we have divided our freight business into a number of main processes. Each process has an owner who is responsible for the continual development of their process. Most processes are supported by Information Systems (IS).

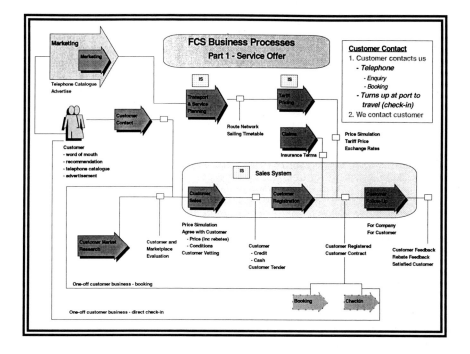

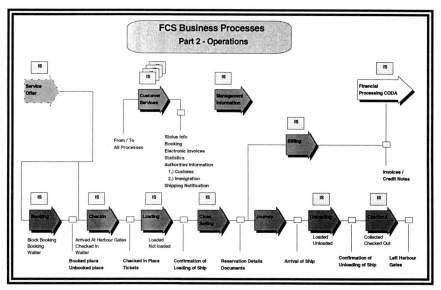

The operational processes can be compared to a conveyer belt, with just a change of status and maybe some additional details at certain points. However, this works only as long as we can obtain as much booking information as possible in advance of the check-in process.

A new IT project is under way to build a new freight system which will allow each IS to run independently of the others with clear interfaces in and out. Different techniques can be used for each IS which best suit the business needs. For example, a graphical based client-server application best serves customer sales allowing our sales personnel to work remotely with the customer and immediately download contract rates into the server when connected online via a GSM mobile telephone.

Alternatively, other IS's may be character based and others it is desirable to automate as much as possible (e.g. loading using handheld scanner technology and later transponder technology).

From a simplistic point of view we have chosen to separate each IS into three basic layers with the user interface being as thin as possible.

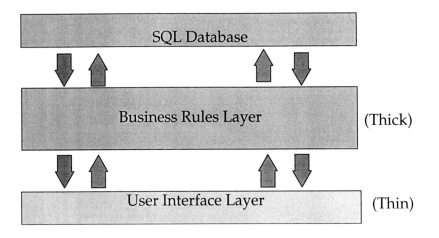

The reason for this is that it means that we create a flexible and open environment which allows us to replace our hardware, relational SQL database, replace different types of technology (character, graphical, semi-automatic (scanners), automatic (transponder techniques)) due to changes in price, performance and direction within the IT marketplace. This in return protects Stena's investment in their future IT solutions.

The following diagram shows our chosen approach to building the IS's.

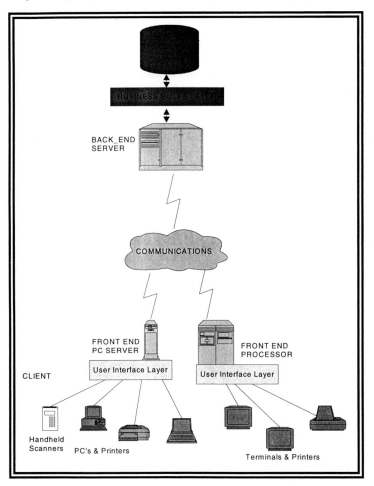

20

FULLY AUTOMATIC SCHEDULING METHODS

Dag Wedelin

ABSTRACT

With the technology available today, fully automatic, optimization based, scheduling methods can be used to significantly improve the efficiency of a wide range of transportation and scheduling tasks. In this paper we will discuss and highlight some design issues for such systems. In particular we will describe our own experience in the development and use of the Carmen system for airline crew scheduling, used by several major European airlines.

Keywords: Algorithms, automatic scheduling, airline crew scheduling

INTRODUCTION

Fully automatic scheduling is fundamentally different from computer aided decision support which merely aims at facilitating manual planning and scheduling tasks. Automatic scheduling relies heavily on mathematical optimization methods and algorithms and can, if well applied, significantly improve the efficiency of transportation and scheduling tasks. In this paper we will discuss and highlight some design issues for such systems. As a case study, we will describe our own experience in the development of the Carmen system for airline crew scheduling, used by several major European airlines.

The systematic study of optimization methods for scheduling in general has a long history, and there are many known and well studied special problems. However, the success of such methods is nontrivial for several reasons. To be introduced at all, there must be an awareness that a system of this kind could lead to a significant increase in productivity, and the IT infrastructure must give access to complete and reliable data about the problem to be solved. Also, in most cases it will not be sufficient to supply only a scheduling algorithm, but it is necessary to have a reasonably complete system including a graphical user interface with manual editing and scheduling, presentation of results, and data import and export functions.

When it comes to modelling, some aspects of real world problems can be difficult to describe in a mathematically desirable form, and may lead to very large optimization problems that are difficult to solve. For some problems, typically where issues such as "fairness" are important, it will be necessary to introduce a (preferably minimum) number of subjective parameters, and extensive empirical calibration may be required before reasonable schedules are obtained.

The most important modelling aspect is to provide a scientifically sound and mathematically clear model, with a conceptually modular separation of problem definition and algorithms. The problem should be defined without unnecessary reference to specific algorithms that may be used for actually solving the problem, and in principle this definition should allow evaluation of any given externally or manually created schedule. This enables a design where it is possible to select different algorithm alternatives for the same problem, and to modify the definition of the problem with minimum changes in the algorithms.

However, the most limiting fundamental factor in the design of automatic scheduling systems are the major algorithmic difficulties. Typically, scheduling involves the management of indivisible entities such as vehicles, people etc, giving the problems a very combinatorial character, which is very difficult to handle. For this reason, many basic difficulties and possibilities of automatic scheduling can be explained by understanding these algorithmic possibilities and limitations.

ALGORITHMIC TECHNIQUES

Generally, an algorithm is considered to be a programmable method to solve a certain class of problems, such as for example multiplication. A particular problem, e.g. 4*7, is an instance of this problem. One of the fundamental properties of algorithms for problems of any complexity is that their execution time grows faster than the size of the problem instances. In Figure 1 we show the typical relation between the size n and the running time $T(n)$ for two different algorithms for the same problem.

An important difference between different algorithms however, is how fast they grow with the size of the problem. The most important distinction is here between polynomial and exponential algorithms. For example, in Table 1 we show the difference between two different complexity functions, the polynomial $10^5 n^2$ and the exponential $10^6 2^n$, where n denotes the size of the problem instance. From the table we can clearly see that with an exponential algorithm, it is hardly possible to solve a large problem, and that buying a faster computer will hardly help at all.

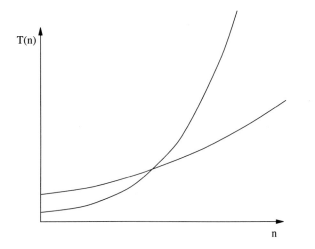

Figure 1. Different relations between problem size (n) and running time (T(n))

Table 1. Comparison between polynomial and exponential algorithm

	10	20	30	40	50	60
n^2	0.001s	0.004s	0.009s	0.0016s	0.025s	0.036s
2^n	0.001s	1s	18 min	13 days	36 years	36600 years

For many simple and basic problems, such as for example the shortest path problem, fast polynomial algorithms are readily available. However, for many other problems, including most discrete optimization problems, there seem to be no such algorithms (they belong to a class of problems known as NP-hard, and can be considered to be intrinsically "slow"). The only algorithmic possibilities that seem to remain are to solve only small problems, or if the problems are large, solve them approximately with some more heuristic approach.

An example of such a difficult problem is the well known and much studied travelling salesman problem, TSP. This problem contains many basic aspects that hold for scheduling in general, and especially for transportation. The task is to find, given a number of cities, the shortest round trip between these. The difficulty is due to the combinatorial explosion, where the number of possible routes increases as (n-1)!, where n is the numer of cities, and the obvious approach to solving the problem by testing all possibilities is intractable except for small problems.

With the TSP as an example, we will now review the main algorithmic approaches to discrete optimization problems of this kind:

- Special cases. Sometimes particular instances of the general problem can be solved. If for example all cities happen to lie evenly along a circle, the best solution is not difficult to find regardless of the size of the problem.

- Enumeration. This is the fundamental search technique of testing cases. To be useful in practice, search must be reduced in some way. The most popular technique here is branch and bound. The idea is to keep the best solution found so far, and avoid searching parts of the search space for which it can be determined that all cases are inferior to this solution. While perfect in theory, it may be too slow to be practical. (In this overview, we also consider dynamic programming as another kind of intelligent enumeration).

- Greedy algorithms. This is a fast way of obtaining a solution, usually of low quality. The idea is to choose whatever looks best in each step, and never change your mind. For example, start at an arbitrary city, and repeatedly move to the closest not visited city.

- Local search. The idea is to repeatedly improve an existing solution, by small changes. For example it may be possible to reverse the order of a subsection of the cycle, thereby effectively eliminating that two links cross over each other. Local search can be quite fast, but there is no guarantee that an optimal solution is found.

- Relaxation techniques. The idea is to relax some difficult requirement in the problem to obtain a simpler problem (very often a linear programming problem), which may give you useful information about the problem you really want to solve. Common relaxations are to skip difficult constraints (such as that we want all cities in a single cycle), or to relax indivisibility (making it possible for the salesman to split in two parts that travel in different directions). If you are lucky, the solution to this simple problem may not use the extra possiblities, and you will find yourself with a solution to your original problem. This technique is often not used by itself, but to obtain bounds to speed up a search process. Depending on the character of individual problem instances, this may or may not be successful.

The choice of algorithms is highly problem dependent. Also, it is typically necessary to make a fundamental compromise between the speed of the algorithms and the quality of the obtained solutions in terms of how well the objective function is maximized or minimized. Special cases, enumeration and relaxation techniques can be used to find truly optimal solutions, while greedy algorithms and local search generally cannot. An optimal approach undisputably gives the best solutions if it succeeds, but may also give unpredictable and unreasonable running times, since they challenge the fundamental difficulty of these problems. Well designed however, a possibly non-optimal design may be preferable, giving other advantages such as reasonable and predictable running times, and a greater flexibility in both modelling and algorithm design.

AIRLINE CREW PAIRING

As a case study we will now consider the airline crew pairing problem, which is a good and representative example of a large real world scheduling problem. Solving this problem is central for managing the crew resources of any airline, the costs of which are the second largest operating cost after fuel.

For a given timetable of flight legs, where a leg is a non-stop flight, the crew pairing problem is the problem of assigning crews to the legs in a fleet. The schedule of a single crew is described by a pairing, which is a chain of flight legs starting and ending at a crew base, and which can be from one up to several days long. The problem consists of constructing a set of pairings such that:

- all flight legs are covered with (at least) one crew;
- all pairings follow legal rules and regulations;
- crew cost is minimized.

As a very simple example consider the following problem adapted from Andersson et al. (1996). This problem is a daily problem, where it is assumed that the timetable consists of six legs. The basic data is given in Table 2.

Table 2. A simple example problem

Leg no	from	to	dep	arr
1	A	B	6.30	13.30
2	B	A	14.30	21.30
3	B	C	10.15	11.45
4	C	B	12.15	13.45
5	B	C	14.15	15.45
6	C	B	16.15	17.45

For a legal pairing, we assume that the following rules apply:

- A pairing must begin and end in either A or B;
- A pairing must be at most two duty periods (days);
- The working time of any day, from the departure of the first flight to the arrival of the last, must not exceed 12 hours;
- A night-stop between two days must be at least 9 hours, or 1.5 times the duty time of the previous day, if this is larger. A night-stop may not exceed 32 hours.

Examples of pairings that are legal are 1-ns-2, 3-4 and 3-4-5-ns-4-5-6, where ns indicates a night-stop. Assuming that the crews are paid a fixed salary, the objective will be to minimize the number of crew days, so a one day pairing will have the cost 1 and a two day pairing the cost 2. Note that to operate all flights every day, two crews are needed for any two day pairing, one starting on odd days, and the second starting on even days. One way to solve this example is to select the pairings 1-ns-2 and 3-4-5-6. These pairings cover all legs, and give a total cost of 3, so three crews are needed to operate this schedule every day.

The main modelling difficulty of this problem is the very complicated rules for legal pairings, which are always much more complicated than in our simple example. These rules are difficult to express mathematically, and even more difficult to use directly in an optimization algorithm. The pairing problem is therefore commonly solved in two steps, with a technique which is generally useful for many kinds of scheduling problems:

1. Firstly, a large number of pairings are generated in an enumerative or similar search process. In principle, every generated pairing is checked against the rules, so that only legal pairings are output from this process. The number of possible pairings can be very large, so even if many pairings are generated, they will usually only be a small fraction of all possible legal pairings. For each pairing a crew cost is calculated based on salaries, additional night-stop costs etc.

2. Secondly, an optimization problem is solved, in which a small subset of the pairings are selected to minimize cost, given the remaining constraints that each flight leg should be covered. Since many pairings have been generated, this problem is usually very large. On the other hand it does not contain any difficult legality constraints, since they were already taken care of in the first step.

The optimization problem of the second step is known as a set partitioning or set covering problem, and can be expessed as a 0-1 integer programming problem in the following way. For each pairing a binary decision variable is defined, and the cost for the variable is the cost of the pairing. Then for each leg a constraint is added, which states that at least one of the pairings with this leg must be selected in the solution. For example, if leg 1 is included in the generated pairings number 2, 5, 13 and 15, the first constraint would be

$$x_2 + x_5 + x_{13} + x_{15} \geq 1$$

The reason that the constraint is not necessarily a strict equality is that it may in some cases be acceptable that a crew travels as passengers, if it turns out that the crew is better needed somewhere else. Using standard mathematical notation, the set covering problem can then be written as

$$\min cx$$
$$\text{subject to } Ax \geq 1$$
$$x \text{ binary}$$

If strict equality is required, the problem becomes a set partitioning problem. It is common that this model is also extended with so called capacity constraints for crew bases, if only a limited number of crews may be chosen from any base.

CARMEN APC

We will now consider how this basic technique is applied in the Carmen system. The Carmen system comprises a complete set of graphical tools for editing pairings and rules, as well as fully automatic generation of complete optimized schedules. Focusing on the automatic pairing generation, the Carmen system follows the general principle in the previous section. Figure 2 gives an overview of the Carmen APC (automatic pairing construction) module, the components of which are described below.

flight and crew data

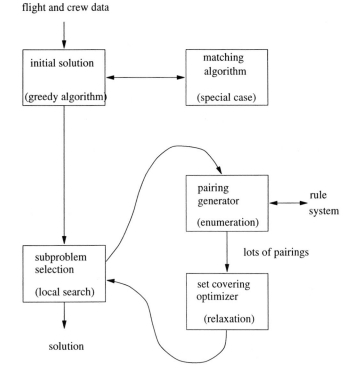

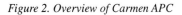

Figure 2. Overview of Carmen APC

A common size for a large weekly problem is around 4000 legs/week, and with typically 3-5 legs per day in a pairing, this would give an intractable number of pairings to enumerate in the generator. However, given the daily pattern with a lot of traffic during the day and little at night, it is a reasonable heuristic to solve daily subproblems in an overall local search pattern. For this purpose there is an overall subproblem selection process working in the following way.

First, an initial solution is created by creating a schedule from left to right, adding one day at a time using the ordinary generator but with other parameters as for the subsequent reoptimization process. To facilitate this process, an optimal matching of all incoming and outgoing flights at each airport is computed, and used as a heuristic guideline.

In this solution a daily window is then opened up as a subproblem and is optimized in the hope of finding an improvement. To optimize the whole problem with this technique, each day is optimized in turn, creating small improvements until nothing more happens. Typically, the days will be traversed 3-5 times.

A difficult and highly pragmatic issue adding to the complexity of defining the subproblems, is the problem of deadheading. Unfortunately, it is usually not sufficient to use only the flights to be scheduled as connections, but passive transfers of crews, or so called deadheads, on other flights are necessary to obtain solutions of high quality. A limited number of additional flights must therefore be added to the chosen subproblem solely for this purpose.

The basic and conceptually simple algorithm for generation is a depth-first search in a search tree determined by the connection matrix representing all legal connections between legs. Search always begins from the subset of legs known as start legs, being legs directly from a crew base, or connecting to a fixed part of a pairing (outside of the subproblem selected) from such a base. Analogously, stop legs leading back to the crew base can be identified.

An important and unique feature of the Carmen system is that the usually complicated rules for feasible pairings are not hardwired into the system, but can be input as any other data, and be written and maintained by the airlines themselves. The rules are entered in an application specific rule language to a separate module, the rule compiler. This module will compile or interpret the rules, and provide the system with a single interface for testing whether a given pairing, or part of a pairing, is feasible or not. Also the cost function for optimization is entered in this way.

To make the generator reasonably efficient, the depth-first search is restricted in several different ways:

- The search is limited by a maximum branching width on the search tree. Typically this width is around 5-8 connections. It is possible to indicate a preference for which connection or connections should be searched first, in order to increase the chance that good pairings will in fact be found within the limited search width. One empirically useful way of finding preferred connections is to follow the aircraft in a connection, if this is known.

- In every step of the depth-first search, the pairing built so far is checked for legality. If it is illegal, this branch is not investigated further. It is important to note that since the legality rules are programmable by the end user, they can also be used to actively enforce not only genuine legality rules but also additional rules for the purpose of pruning the search. For specific problems, an experienced user will often be able to know that some connections are useless etc., and this information can be very useful.

The set covering problem is a special case of the more general 0-1 integer programming problem, for which many methods of solution are known. The most common exact method is to use a branch and bound technique, with a linear programming relaxation for the bound. However, this method, and any other known method for solving these problems suffer from the exponential growth of combinations, and places a limit on the range and size of problems that can be solved. Other methods, such as greedy heuristics, can often give feasible solutions to large problems, but of poor quality.

Unfortunately, since one wants to generate a large number of pairings in the generation step, for the Carmen system typically between 10,000 and 1,000,000 pairings, the optimization problems are very large, and also high quality solutions are clearly necessary. Using standard techniques, the optimizer therefore becomes a bottleneck in the system. In some systems, this is handled by restricting the number of pairings generated, and by using a more complicated optimizing generator. However, it is then very difficult to separate the rules from the generator, which is a significant design advantage of the Carmen system.

As a first and trivial step, preprocessing is applied to reduce the size of problem. However, the problems will still be very large, and to solve this, the Carmen system uses a new and innovative type of optimizing algorithm. This optimizer has been described elsewhere, and we refer to Wedelin (1995) for a full technical description. However, we will here outline the approach. In contrast to most other integer optimizers the optimizer does not use a branch and bound technique, but uses a strategy of manipulating the cost function (i.e. the reduced costs) using the invariants given by the constraints, until the solution can be easily determined. The critical idea of the algorithm is to allow also manipulation of the cost function that is not invariant, so that an integer solution can be found. Mathematically, the optimizer can be viewed as a dual relaxation strategy, and it can be shown that for different parameter settings the algorithm can be interpreted in terms of linear programming, dynamic programming and greedy algorithms.

In practice, it turns out that for set covering, the optimizer can as a result of its design easily and predictably solve very large set covering problems up to about two million variables, the limiting factor being not execution time but memory. With respect to quality, it turns out that the results (for smaller problems that can be compared) are of the same quality as from optimizers such as CPLEX, based on linear programming and branch and bound.

For the entire system, the quality and running time depend on many factors in a complex way. For many problems the system appears to produce near-optimal solutions, although this cannot be proved. Perhaps the most important result is that the solutions are usually better than plans made by manual planners, and that in several cases, the pairing productivity has been increased by around 10% in the planning phase. The running time is affected by many internal parameters that are not a direct function of the problem itself, such as the search width of the generator, the number of generated pairings etc. In practice, these parameters are set to make the system give a solution in a reasonable time, say in an over-night run.

CONCLUSION

We have considered a number of design issues for automatic scheduling methods. In the modelling step the definition of the problem should not be confused with the algorithm to solve it, although it is also necessary to have a firm grasp of the possible algorithmic alternatives. Also, we have discussed various algorithmic techniques of avoiding the combinatorial explosion, and shown that in the Carmen system all of these techniques are used together to create a complete automatic scheduling system.

A reason for the success of this project in an area, where many others have failed, is that there has been a successful integration of state of the art research, development of a complete system with integration of manual and automatic planning tools. Another is the technological innovations used in the Carmen system such as the rule compiler making it possible to handle and customize the complicated rules characteristic for European airlines, and the optimizer capable of solving very large problems.

REFERENCES

Andersson E., Housos E., Kohl N., Wedelin D. (1996) Crew Pairing Optimization. In Yu: OR in Airline Industry.

Desrosiers J., Dumas Y., Solomon M., Soumis F. (1995) Time constrained routing and scheduling. In Handbooks of Operations Research and Management Science vol 8: Network routing, North Holland.

Gershkoff I. (1989) Optimizing Flight Crew Schedules, Interfaces 19-4, 29-43.

Wedelin D. (1995). An algorithm for large scale 0-1 integer programming with application to airline crew scheduling. Annals of Operations Research 57, 283-301.

Wedelin D. (1995). The design of a 0-1 integer optimizer and its application in the Carmen system. European Journal of Operational Research 87, 722-730.

21

OPTIMISATION OF A GENERIC MULTI-NODE TRANSPORTATION AND INFORMATION NETWORK

Stig Andersson
Karl Jansen
Jonas Waidringer

ABSTRACT

In this paper we are going to briefly describe a model for transportation that is characterised by four features. Firstly the nodes are non-stationary and can be adapted to accommodate changed requirements on the transportation system. Secondly a high degree of complexity is built into the operational modes of the nodes. Thirdly, the model incorporates an information network interacting closely with nodes and links. Finally, all networks, systems etc. will be optimised employing very powerful algorithms and tools from system theory and combinatorial graph theory. These features make the model both flexible and realistic.

The model aims at solving a type of problem where, as an effect of for example external influence, nodes and hence also links may be changed and transformed in a dynamic fashion. The model obviously, and in a quite generic way, includes a vast number of logistic networks and situations, ranging from container routing, complex delivery systems to highly optimised military logistic schemes. Basically the model works with flows on graphs, or specifically networks, where the nodes are characterised by a number of parameters describing the various functionalities of the node.

Keywords: Neural networks, combinatorial graph theory, transport models

THE INTERACTION OF TRANSPORTATION AND INFORMATION NETWORKS

In transportation it is important that the information associated with goods or passengers is correct and arrives at the proper moment. Great amounts of resources have been invested into information systems to increase the efficiency of goods handling and passenger transport. However, to determine if these solutions are optimal is harder, but certainly worthwhile to investigate. An investigation of this kind needs a structured model in which an optimisation is well-defined and can be carried out.

To describe the modelling problem conceptually we can for example consider a small port that has a few nodes where goods are reloaded or stored and, associated with the transportation network, an information network. A larger system can be described using the same concept; however with an increasing number of nodes in the system, the computational complexity increases exponentially. This means that the scenario's that are to be modelled have to be considered very carefully.

First it is important to point out that physical flows and information are two parts of a larger network, since the interaction is very strong and important for the system to work. When considering a transportation system it is soon noticed that information is not only generated by documents, EDI or other data specially prepared for the transport. Information also originates from the transport system itself; for example all goods have to be present for a delivery to be carried out. Another example is some storage area that has reached its maximum capacity; goods will have to be re-routed and someone has to be notified. Information from outside the system can affect the system as well. From a modelling point of view all different kinds of information flows are described in the same way.

Within the system there are two general types of information available, *static* and *dynamic*. The static information supplies the framework for the system studied. During the time of the modelling this information is constant. The dynamic information changes during the modelling and includes among other things documents associated with the transport. This type of information can be more or less random. The longer the time period over which the simulation is carried out the more information is of the dynamic type. Since the dynamic information has different time cycles, some of which are short while others are longer, it is possible to divide the modelling into smaller time windows where some of the information can be considered to be static.

In this integrated information and transportation network the information can be modelled in a fashion similar to a chemical reaction. However a large difference from a chemical reaction is that stoichiometry and equilibria are not included in general.

The simplest case is:

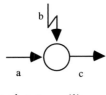

$$a + b \rightarrow c \qquad (1)$$

where a and c are goods flows and b is a binary "flow" regulating the throughput from a to c. If b=0 then there is no flow allowed from a to c, and if b=1 the flow is turned on. This example may be considered as a binary valve. In a more complicated case b could supply information about how much flow is allowed between a and c at a given moment. An example that is more realistic for transportation may look like this:

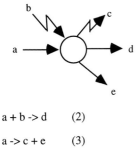

a + b -> d   (2)

a -> c + e   (3)

In this example if the information, for example a loading order, supplied by flow b is present then the goods will be routed along d. However, if the information b, the loading order, is missing then the goods will be routed along e, for example a special storage area, and information about this is sent with the flow c, to a shipping agent. From the examples above it can be inferred that to be able to control and apply efficient strategies for an adaptive network of this type, informatics support is indispensable. In some cases extremely fast support of this type is needed. The model is therefore based on an associated database, where the pieces of information, "statements", have been graded in relevance with respect to all potential questions which could be asked within the framework of the model. This can be a serious problem in some cases since transportation systems have not been analysed and considered from this point of view before.

Information and personnel flows are considered in the manner described above and are incorporated into the model as mathematical equations at the nodes. There is another approach - micro simulation - where every single goods unit is included with its own equations. However, micro simulation is much more complex and requires the flow model as an input.

Also as mentioned above, the model will optimise a foliated network, where flow network, information network and set of resources will be simultaneously optimised. In modelling, it is very useful to use dynamic points of view, where the links by no means always connect all nodes but instead a link - and hence the whole network - is established only at the moment when a unit from the resource set is allocated to serve between two given nodes.

INCLUSION OF COMPLEX NODE FUNCTIONALITIES

In standard combinatorial graph theory the nodes are intersections between links, the only thing that happens at a node is that the flow is routed along one link or another. Everything that happens in a node happens momentarily. For electric circuits this is true but not for transportation flows. To model transportation flows more correctly the nodes, not just the links have to be given properties. Properties that can be assigned to the nodes are storage capacities, turn over rates and so on. The nodes can be more complex than just the information functions described above. These properties correspond to those given to the links. It is the basic ideas of neural networks that are used to model the complex and flexible nodes. Included in neural network theory are also powerful algorithms for parallel computation which are highly useful for modelling transportation. Neural networks *are never programmed; they learn.* The characteristics and flexibility of the nodes are given by mathematical functions incorporated in the nodes, functions that can differ significantly from node to node. One of the simplest functions that can be assigned to a node is the step function, which at a given time switches from zero to one or vice versa. There are no theoretical limits to which functions may be used; computer capacity and imagination set the limits.

The connections between the nodes are modelled with combinatorial graph theory. Benefits of this theory are that not only the links, but also weights, between the nodes are included and may be easily varied. These weights can have the meaning of capacities, transportation times and other parameters describing the interaction between nodes. The state variables that one uses to build an algorithmic theory for the nodes have meanings that correspond to those of the weights between nodes. For example the turn over rate for the node corresponds to transportation times and storage capacity corresponds to some capacity of the link. All node variables are of course functions of other parameters.

Parameters that can be included in the nodes are:

- storage capacities of the nodes, related to the quality of goods, size of warehouses and other storage areas;
- link capacities, related to the amount of vehicles available, quality of roads, etc.;
- flexibility, how easy it is to change the node from flexible to rigid;
- thresholds, when a limit is reached something happens, e.g. a full container is shipped away.

OPTIMISATION

In mathematical terms the model is quite complex, since the degrees of freedom are directly coupled to the number of links and nodes in the system as well as to the parameters within the links and nodes in a quite refined way. The number of these parameters is very large in a real transportation system. To be able to optimise complex dynamical systems of this kind powerful algorithms from combinatorial graph theory and neural networks theory as well as ideas and concepts from several other areas of science have to be used. With these concepts and algorithms well at hand, several typical optimisation and simulation issues on transportation networks can be addressed. One of these issues is optimal flow problems, where capacity functions in both links and nodes are basic entities. Using this approach the optimal flow over the network studied is analysed. In the analysis bottlenecks are identified, optimised handling operations and robust realisations are suggested. Another class of problems which is similar to the ones mentioned above is optimal routing problems. In these cases, instead of capacities, cost functions set the boundaries on for example route length, travelling time and energy consumption. Here the aim is to find the best route for the goods or passengers to travel. The third issue deals with optimal strategies to minimise the influence of removal of links and nodes from the system. This question is called system stability.

In all three cases described above information flows and spontaneous nodes are incorporated into the optimisation process. To make an optimisation the essential parameters of the studied system have to be identified, these may not always be the ones that seem obvious. In fact a bottleneck seen at one location might be the effect of problems in completely different places of the network. Only a qualified analysis of the network will be able to identify the problems.

Several necessary steps have to be taken to perform any of the above mentioned optimisations. First the transport system studied has to be analysed and input data, together with specifications of nodes and links must be collected. This work requires much time since needed data has usually not been presented in a way suitable for modelling before. The next step involves constructing the physical graph. This is a first step in network synthesis. When this is done the optimisation problem for the network is transformed into a problem in matrix form. There are several matrices used to describe the system (connectivity matrix, capacity matrix, reachability matrix etc.).With the proper algorithmic procedures the matrix problem is solved. In most cases there is a certain parameter, for example travel time or cost, that has to be minimised. In most cases other variables define the boundary conditions. Before the solution can be accepted the optimisation and stability properties of the system have to be verified.

NETWORK SYNTHESIS

Conversely, one may also study the network synthesis question, i.e. for a given set of flow requirements, construct a minimal physical network implementing these requirements. Results in this direction are roughly speaking of two kinds: existence proofs and constructive procedures. An existence proof will tell if a certain set of requirements can at all be physically implemented. This is of course of paramount importance to know before even worrying how to do it. A constructive procedure will provide directions how to implement, if possible, a given set of flow requirements.

REFERENCES

Batten, D., Casti, J. and Thord, R. (Eds.), "Networks in Action", Springer-Verlag, 1995

Evans, J. R. And Minieka, E., "Optimisation Algorithms for Networks and Graphs", 2nd edition, Marcel Dekker, Inc., New York, 1992

Huckenbeck, U., "Extremal Paths in Graphs", Akademie Verlag, Berlin 1997

Jungnickel, D., "Graphen, Netzwerke und Algorithmen", 3rd edition, Wissenschaftsverlag, Mannheim, 1994

Taylor, J. G., Caianiello, E. R., Cotterill, R. M. J. and Clark, J. W. (Eds.), "Neural Network Dynamics", Springer-Verlag, 1992

ABOUT THE AUTHORS

Andersson, Stig, is Research Director of CECIL, Center at Eriksberg for Communication, Information and Logistics, Gothenburg.

Bergendahl, Göran, is professor of business administration, Department of Business Administration, School of Economics and Commercial Law, Gothenburg University.

Dubois, Anna, is a doctoral student at the Department of Systems Management, School of Technology Management and Economics, Chalmers University of Technology, Gothenburg.

Edwards, Henrik, is a doctoral student at the Department of Business Administration, School of Economics and Commercial Law, Gothenburg University.

Forsström, Åke, is professor of economic geography, Department of Human and Economic Geography, School of Economics and Commercial Law, Gothenburg University.

Gadde, Lars-Erik, is professor of industrial marketing, Department of Systems Management, School of Technology Management and Economics, Chalmers University of Technology, Gothenburg.

Hellgren, Johan, is a doctoral student at the Department of Transportation and Logistics, School of Technology Management and Economics, Chalmers University of Technology.

Hoek, Geert, is an industrial engineer of the Faculty of Technology Management, Eindhoven University of Technology, Eindhoven, Netherlands.

Högberg, Kjell, is an information systems manager at Bilspedition Traffic & Information Systems, Division Europe, Gothenburg.

Hultkrantz, Ola, is a doctoral student at the Department of Transportation and Logistics, School of Technology Management and Economics, Chalmers University of Technology.

Ireståhl, Bo, is Executive Director of Bilspedition Traffic & Information Systems, Division Europe, Gothenburg.

Jansen, Karl, is a research associate at CECIL, Center at Eriksberg for Communication, Information and Logistics, Gothenburg.

Jensen, Arne, is professor of business administration, Department of Business Administration, School of Economics and Commercial Law, Gothenburg University.

Kanflo, Thomas, is a doctoral student at the Department of Transportation and Logistics, School of Technology Management and Economics, Chalmers University of Technology, and staff member at Bilspedition Traffic & Information Systems, Division Europe, Gothenburg.

Larsson, Anders, is a doctoral student at the Department of Human and Economic Geography, School of Economics and Commercial Law, Gothenburg University.

Lindeberg, Henrik, is a logistics manager at SKF Distribution AB, Gothenburg.

Lonntorp, Johan, is head of Control and Systems Support, Cargo Division, Stena Line Service AB, Gothenburg.

Lumsden, Kenth, is associate professor of transportation and logistics, Department of Transportation and Logistics, School of Technology Management and Economics, Chalmers University of Technology.

Polewa, Rudolf, is a doctoral student at the Department of Transportation and Logistics, School of Technology Management and Economics, Chalmers University of Technology, Gothenburg.

Samuelsson, Anders, is a doctoral student at the Department of Human and Economic Geography, School of Economics and Commercial Law, Gothenburg University.

Sjögren, Stefan, is a doctoral student at the Department of Business Administration, School of Economics and Commercial Law, Gothenburg University.

Sjöstedt, Lars, is professor of transportation and logistics, Department of Transportation and Logistics, School of Technology Management and Economics, Chalmers University of Technology, and chairman of the Board of the Center for Transport and Traffic, Chalmers University of Technology and Gothenburg University.

Svahn, Peter, is a doctoral student at the Department of Business Administration, School of Economics and Commercial Law, Gothenburg University.

Svensson, Rune, is former President, Volvo Transport Corporation, Gothenburg.

Tilanus, Bernhard, is professor of quantitative economic methods, Graduate School of Industrial Engineering and Management Science, Eindhoven University of Technology, Netherlands, and Erik Malmsten professor at School of Economics and Commercial Law, Gothenburg University.

Waidringer, Jonas, is a research associate at CECIL, Center at Eriksberg for Communication, Information and Logistics, Gothenburg.

Wedelin, Dag, is a consultant at Carmen Systems AB, Gothenburg, and staff member at the Department of Computer Sciences, Chalmers University of Technology, Gothenburg.

Woxenius, Johan, is a doctoral student at the Department of Transportation and Logistics, School of Technology Management and Economics, Chalmers University of Technology, Gothenburg.

INDEX OF KEYWORDS